Polymers
Properties and Applications

12

Lev Zlatkevich

Radiothermoluminescence and Transitions in Polymers

With 60 Illustrations

Springer-Verlag
New York Berlin Heidelberg
London Paris Tokyo

Lev Zlatkevich
Department of Materials Science and Engineering
The Technological Institute
Northwestern University
Evanston, IL 60201
U.S.A.

Library of Congress Cataloging in Publication Data
Zlatkevich, L. (Lev)
  Radiothermoluminescence and transitions in
polymers.
  (Polymers, properties and applications; 12)
  Includes bibliographies and index.
  1. Polymers and polymerization—Thermal
properties.   2. Polymers and polymerization—
Radiation effects.   3. Thermoluminescence.   I. Title.
II. Series.
QD381.9.T54Z53   1987   547.7′045   86-27928

Typeset by The Maple-Vail Book Manufacturing Group, Inc., Binghamton, New York.
Printed and bound by Quinn-Woodbine Inc., Woodbine, New Jersey.
Printed in the United States of America.

9 8 7 6 5 4 3 2 1

ISBN 0-387-96407-X Springer-Verlag New York Berlin Heidelberg
ISBN 3-540-96407-X Springer-Verlag Berlin Heidelberg New York

In memory of my grandmother, Roza

Preface

This book deals with one of the recently developed methods of the analysis of temperature transitions in polymers—the radiothermoluminescence method.

Although thermoluminescence from irradiated inorganic materials was first found and examined as far back as the beginning of this century, a systematic study of this phenomenon in organic solids was initiated considerably later, in the 1960s. In the past 25 years, essential achievements have been made both in understanding the mechanism of radiothermoluminescence in organic substances and in practical applications of the technique.

The results obtained to date are described and discussed in relation to our knowledge about structure and temperature transitions in polymeric systems.

Lev Zlatkevich

Contents

Chapter 1

Luminescence

1.1 Luminescence as a Phenomenon

Light is a form of energy, and in accordance with the fundamental concept of energy conservation, energy must be supplied to every material system emitting light. There are two processes by which the material can become a generator of light after absorbing suitable energy [1]. In one process, the absorbed energy is converted into heat. The thermal agitation of all molecules within the system increases, and simultaneously, more and more of the molecules are transfered into excited states. The higher the temperature, the greater is the number of excited molecules and the greater is the intensity of the emitted light. In the other process, the molecules are brought into excited states without increasing their average kinetic energy and without heating the system. An appreciable part of the absorbed energy is temporarily localized as excitation of atoms or small groups of atoms which then emit light; this process is called *luminescence*. Luminescence is characterised by emission of light in excess of the thermal radiation produced by heat in a given material. The basic rule for distinguishing between thermal and light radiations can be formulated as follows: If the intensity of the emitted light exceeds the intensity of the radiation of the same wavelength from a black body of the same temperature, the radiation is a case of luminescence [2].

Luminescence occurs when an excited molecule returns to the ground state with the emission of a quantum of light. There are many ways by which excited molecules are produced, and where luminescence is subsequently observed, the mode of excitation is often referred to in the term used to describe the phenomenon. For example, *photoluminescence* implies the excitation by prior absorption of light, whereas *radioluminescence* presumes the excitation by irradiation with X-rays, γ-rays, electrons, or fast particles. In addition, in such phenomena as triboluminescence and sonoluminescence, excitation is accomplished by shock waves, and in chemiluminescence and bioluminescence, the emission of light betrays the existence of excited molecules produced by a chemical process. The appearance of luminescence, in fact, always implies the presence of excited molecules. Consequently, luminescence is a valuable tool for studying the chemistry of excited states, and since all chemical reactions involve excited states of some kind or another, luminescence in its broadest sense illuminates the whole field of chemistry [3]. This is not to imply that all excited states luminesce. As a general rule, luminescence increases in efficiency as the motion of a molecule is restricted, since the competing processes of radiationless energy transfer require coupling between the excited molecule and the molecules which surround it and this coupling becomes greater with increased amplitude and diversity of molecular motions. Lowering the temperature, therefore, usually increases the probability of luminescent processes. Since the process of luminescence is the de-excitation of excited molecules by re-emission of absorbed (or otherwise obtained) energy as light quanta, it is in direct competition with chemical reactions and every quantum emitted is wasted from the chemical point of view. However, the way in which luminescence changes in different media or with temperature or is quenched or enhanced by the addition of other molecules or otherwise can tell much about the processes of energy transfer and the chemical reactions which may be taking place.

1.2 Electronic Excited States

The internal energy of a monatomic molecule is defined exclusively by the configuration of its electrons. The corresponding energy levels of the atom are in general separated from each other by relatively large intervals. In diatomic and polyatomic molecules, energy is also contained in the vibrations of the atomic nuclei relative to their center of gravity and in the rotation of the molecule around the main axis of inertia. The spacings between the corresponding energy levels which are superimposed on the electronic levels are much narrower than those between the electronic levels themselves.

The total energy of a given state is the sum of electronic (E_e), vibrational (E_v), and rotational (E_r) energies:

$$E = E_e + E_v + E_r \qquad E_e > E_v \gg E_r$$

In the normal configuration of most molecules, the shared electrons in a given bond between atoms are paired with regard to electron spin, and the spin effects (resultant spin) cancel. The multiplicity of a state with regard to the resultant spin s is $2s + 1$. This implies that there are $2s + 1$ ways in which the resultant spin can couple with the orbital angular momentum along the molecular axis to yield the total angular momentum along that axis. Thus, when the electrons are paired, the resultant spin is 0, and the multiplicity of the state is 1. This is the singlet state. In a triplet state, the electrons are unpaired, the resultant spin is 1, and the multiplicity is 3.

The electronic excited states and the most important processes occurring in these states are shown in Fig. 1.1. The basic ground state S_0 is the initial state of unexcited molecule. In excited singlet states, S_1, S_2, etc. electron spins are antiparallel. Most reactions occur between the first excited singlet S_1 and ground-state singlet S_0. This is so because of the very high velocity of transition from higher states S_2, S_3, etc. to lower excited singlet state S_1.

The lowest triplet state T_1 is formed mostly from the lowest excited singlet state S_1. In triplet state, electron spins are parallel and the formation of the triplet state by direct absorption of a photon is forbidden. Higher triplet states T_2, T_3, etc. are formed only from the lowest triplet state T_1 by absorption of another photon. For every excited singlet state of a molecule there is a corresponding triplet state. Because electrons with parallel spin are obliged by the Pauli principle to avoid each other, the electrostatic repulsion between them is less. As a result, a triplet state is always lower in energy than the corresponding singlet state (Hund's rule).

In descriptions of radiation transitions between various electronic states, the symbol of the upper state is always written first, independent of whether this process is absorption or emission. The direction of the transition is indicated by an arrow. Thus processes $T_2 \leftarrow T_1$ and $S_1 \leftarrow S_0$ are light absorptions, and processes $T_1 \rightarrow S_0$ and $S_2 \rightarrow S_0$ are light emissions.

In addition to the radiative processes which result in a visible or otherwise detectable luminescence, there exist a variety of pathways by which an excited molecule may lose an energy which do not involve radiation [4]. These can be classified into two main groups described as internal and external processes. Internal processes of

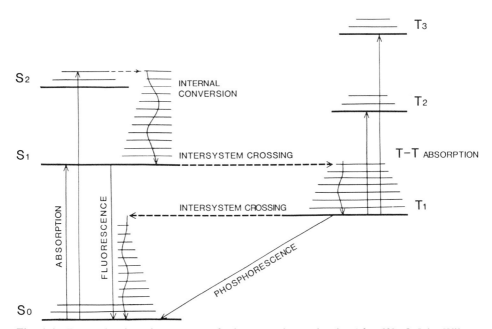

Fig. 1.1. Energy levels and energy transfer in a complex molecule. After [3], © John Wiley & Sons, Inc., with permission.

energy transfer include internal conversion between excited singlet states, which is extremely efficient and results in the lowest exited singlet being the only state which fluoresces, and intersystem crossing to the lowest triplet state, which gives rise to phosphorescence. In each of these processes, the first state is rearrangement of the electron density in the molecule's electron cloud and conversion of electronic energy into vibrational energy. *Internal conversion* is usually taken to mean a nonradiative transition between two states of like multiplicity. When the radiationless transition occurs between states of different multiplicity, the term *intersystem crossing* is used. Radiationless transitions are depict by a wavy arrows, e.g., $S_1 \rightsquigarrow S_0$ and $S_1 \rightsquigarrow T_1$.

In external (intermolecular) processes, a direct transfer of electronic energy to another molecule takes place without the mediation of collisions. For this to occur, some kind of coupling between donor and acceptor is necessary. Depending on the distance separating the two molecules, several mechanisms, such as radiative transfer, resonance transfer, or excitation transfer, are possible.

1.3   Fluorescence and Phosphorescence

The initial stage of the molecule before excitation is usually the singlet ground electronic state S_0. As a result of the emission from electronic excited states, the molecule returns to the ground electronic state, although frequently to a vibrationally excited form of the ground electronic state.

State multiplicities are important because the nature of the emission processes depends on them. If the states from which the emission originates and terminates have the same multiplicity, the emission is called *fluorescence*. This process most commonly occurs between the first excited singlet S_1 and the ground state singlet. If the states from which the emission originates and terminates differ in spin ($\Delta s \geq 1$), the emission is known as *phosphorescence*. The states of principal concern relative to phosphorescence will be the lowest excited triplet state T_1 and ground state singlet S_0. The T_1 state is formed mainly by radiationless intersystem crossing from lower excited singlet state S_1.

The transition probabilities between different energy levels of one and the same molecule may be of quite different orders of magnitude. Great transition probabilities correspond to strong absorption and emission, and vice versa. If the transition probability is extremely small, the transition is called *forbidden*.

The duration of the absorption process is estimated to 10^{-15} sec [3]. The time during which a molecule remains in an excited state before it returns spontaneously, with light emission, to a lower state, is called the *lifetime* of the excited state. Because highly probable transitions can occur only when $\Delta s = 0$, fluorescences have relatively short lifetimes (10^{-7}–10^{-10} sec), whereas phosphorescence emissions have relatively long lifetimes (10^{-3}–10 sec). Consequently, molecules in the triplet state can easily lose their energy by a variety of radiationless processes.

Individually, fluorescence and phosphorescence will normally be the same in all respects, such as wavelength, shape, and lifetime, no matter which excited singlet state is initially occupied. The foregoing occur because the internal conversion process is some 10^3 (or greater) more likely than is emission directly from any excited singlet state above the first excited singlet. Internal conversion from the first excited singlet state can occur, thus competing with and thereby quenching fluorescence and intersystem crossing (therefore, also phosphorescence). However, internal conversion from the first excited singlet state is much less likely than among the higher excited singlets, and thus the complete quenching of fluorescence or phosphorescence is extremely rare.

From a practical viewpoint, the only clear way to distinguish between fluorescence and phosphorescence is to study the effect of temperature on the decay of the luminescence. Fluorescence is essentially independent of temperature, whereas the decay of phosphorescence exhibits a strong temperature dependence.

1.4 Delayed Fluorescence

It is possible that the observed lifetime of fluorescence will be longer than that expected based on "prompt" emission from the S_1 state. *Delayed emission,* sometimes of a much longer lifetime, is observed whenever a process leading to the formation of S_1 or T_1 is kinetically limiting for emission. *Prompt fluorescence* means $S_1 \rightarrow S_0 + h\nu$ sequence. *Delayed fluorescence* may arise if, after a fast intersystem crossing, T_1 is thermally popped back into S_1. Thus, for sequence $S_1 \rightarrow T_1 \rightarrow S_1 \rightarrow S_0 + h\nu$, it may take a "long time" for the emission to occur from S_1. The lifetime of delayed fluorescence is about equal to the lifetime of phosphorescence. This indicates that delayed fluorescence appears as a result of the activation of molecules in the triplet state. On electronic excitation to any state above the first excited singlet, commonly de-excitation occurs to the lowest excited singlet state. From this state there can occur fluorescence emission, continued de-excitation to the ground state with no emission, crossing to the triplet state, or any combinations of these. If the triplet state is occupied, phosphorescence, delayed fluorescence, or de-excitation to the ground state can occur. Because the lowest triplet is below the lowest excited singlet state, phosphorescence occurs at longer wavelength than does fluorescence.

The following is a sequential outline of the processes of concern with approximate lifetimes:

$$S_n \xleftarrow{\quad 10^{-15} \text{ sec} \quad} S_0 \qquad \text{(absorption)}$$

$$S_n \xrightsquigarrow{\quad 10^{-11}-10^{-14} \text{ sec} \quad} S_1 \qquad \text{(internal conversion)}$$

$$
S_1
\begin{cases}
\xrightarrow{\quad 10^{-7}-10^{-9} \text{ sec} \quad} S_0 + h\nu & \text{(fluorescence)} \\[6pt]
\xrightsquigarrow{\quad 10^{-8} \text{ sec} \quad} T_1 & \text{(intersystem crossing)} \\[6pt]
\xrightsquigarrow{\quad 10^{-5}-10^{-7} \text{ sec} \quad} S_0 & \text{(internal conversion)}
\end{cases}
$$

$$\overset{10-10^{-3} \text{ sec}}{\longrightarrow} S_0 + h\nu \qquad \text{(phosphorescence)}$$

$$T_1 \overset{10-10^{-3} \text{ sec}}{\leadsto} S_0 \qquad \text{(intersystem crossing)}$$

$$\overset{10-10^{-3} \text{ sec}}{\leadsto} S_1 \rightarrow S_0 + h\nu \quad \text{(delayed fluorescence)}$$

1.5 Thermoluminescence

Fluorescence is due to the spontaneous transition of a molecule from an excited state to a lower energy level. The mean life of this process depends only on the transition probability and is in most cases very short. Furthermore, it is practically independent of temperature. The characteristic feature of phosphorescence is that excited molecules do not necessarily immediately begin to emit light by returning from the excited state E_k to the ground state E_0 (Fig. 1.2). A fraction of the excited molecules may pass instead into a metastable or quasistable state E' of somewhat smaller energy than E_k. From E', the molecules can only return to E_k, with subsequent light emission accompanying the passage $E_k \rightarrow E_0$ when the missing energy $(E_k - E')$ is restored to them by the heat movement of the surrounding medium [5]. This return occurs within a short time at high temperatures. It might be delayed a good deal, however, at low temperatures. At very low temperatures, the phosphorescence may be completely "frozen in." The molecules then remain in their quasistable state E' until, by a rise of temperature, the "trapped" radiation is again released. The smaller the energy difference between E_k and E', the lower is the temperature necessary for "freezing in" the phosophorescence. With high $(E_k - E')$ values, phosphorescence

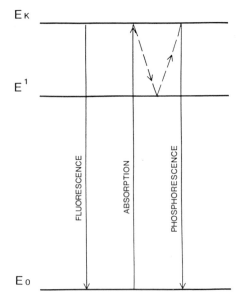

Fig. 1.2. Energy levels for the production of thermoluminescence. After [5], © John Wiley & Sons, Inc., with permission.

may even be frozen in at room temperature. This is the origin of so-called *thermoluminescence*.

1.6 Absorption and Emission Spectra

An *absorption spectrum* is produced when light with a continuous spectrum is passed through a medium which absorbs only part of the light. The quantity related to absorption that is conventionally measured is called the *optical density* (OD) and is defined as being equal to log (I_0/I_t), where I_0 is the intensity of incident light, and I_t is the intensity of transmitted light. That is, an optical density of 2.0 corresponds to ~1% transmittance or ~99% absorbance, and an optical density of 0.01 corresponds to ~98% transmittance or ~2% absorbance [6]. An absorption spectrum is completely described by a graph with absorption intensity as the ordinate and wavelength of absorbed light as the abscissa. Selective light absorption by organic substances depends on their electronic structure, particularly on the presence of electron vacancies in a molecule.

An *emission spectrum* is a plot of emission intensity I_e (at a fixed absorption intensity) as a function of frequency ν or wavelength λ of absorbed light. If a molecule is raised into an excited state E_k by light absorption or any other kind of mechanism, and if, apart from the ground state with energy E_0, several other energy levels E', E'', etc., are situated below E_k, the emission spectrum will result from transitions from E_k to all or some of the levels E_0, E', E'', etc. The levels E', E'', etc. can either correspond to different electronic excitation states or they can be due to the superposition of different vibrational energies on one and the same electronic state. The emission spectrum originating from one single excited state will thus consist of a series of lines. In more complicated molecules, especially in condensed systems (in liquids or solids), the narrowly spaced energy levels become more or less broadened and frequently overlap. Accordingly, sequences of broad bands will appear, and eventually, these will merge into one continuous band showing several peaks of higher intensity.

In condensed systems, interactions with other molecules during collisions or with neighbors in the crystal lead to a removal of part of the excitation energy before reemission can take place. As a result, the emission spectrum always arises from a level lower than that reached by absorption and is in consequence displaced toward longer wavelengths compared to the absorption spectrum.

References

1. Leverenz, H.W.: An Introduction to Luminescence of Solids. Wiley, New York, 1950
2. Fonda, G.R.: J. Appl. Phys. *10*, 408 (1939)
3. Windsor, M.W.: Physics and Chemistry of the Organic Solid State, vol. 2 (ed. Fox, D.). Interscience, New York, 1965
4. Somersall, A.C., Guillet, J.E.: J. Macromol. Sci. *C13*, 135 (1975)
5. Pringsheim, P., Vogel, M.: Luminescence. Interscience, New York, 1946
6. Turro, N.J.: Modern Molecular Photochemistry. Benjamin/Cummings, Menlo Park, Calif., 1978.

Chapter 2

Interaction of Radiation with Matter

The term *high-energy radiation* is applied both to particles moving with high velocity—fast electrons or β-particles, fast protons, neutrons, α-particles, and charged particles of higher mass—and to electromagnetic radiation of short wavelength—X-rays and γ-rays. The processes by which these different forms of radiation react with the atoms of a specimen through which they pass may be different, the common feature being the large amount of energy carried by each particle or photon (this energy is very much greater than that binding any orbital electron to an atomic nucleus or an atom to its neighbor). In this respect, they differ from slow or thermal neutrons and from ultraviolet light, in which the energy carried per particle or photon is usually smaller than the ionizing energy of an atom or molecule. The effects produced by γ-rays or X-rays may be best understood as due to discrete high-energy photons, which may therefore also be considered as particles in this context. In most cases, the changes produced by incident radiation depend mainly on the total energy absorbed and very little on the type of radiation or its energy.

2.1 Energy Absorption and Trace Structure Produced by Charged Particles. Thermolization of Electrons

In passing through matter, all forms of high-energy radiation lose energy by reacting with electrons and nuclei of the medium and may give rise to displaced nuclei, free electrons, ionized atoms or molecules (which have lost their electrons), excited atoms or molecules (in which an electron is raised to a higher energy level), and radicals (uncharged molecules with an unpaired electron). Neutrons affect the nuclei of atoms; other energetic radiations affect primarily the electrons. In molecular structures in which atoms are bound together by shared electrons to form molecules, excitation and ionization are by far the most important effects of exposure to high-energy radiation. The ratio of energies lost in excitation and ionization is not known accurately, and it is generally assumed that they are comparable in amount [1].

For most high-energy radiations, the bulk of the energy is eventually deposited in the medium through the release of fast electrons. The motion of a fast electron through condensed matter is punctuated by a copious number of well-separated excitation events, each of which may constitute either a single or a group of individual molecular processes. In this way, the particle is slowed down and its energy is progressively transferred to the medium.

From the point of view of chemical effects, the rate of energy loss per unit of path travel is of the most importance. This determines the penetration of the incident beam, as well as the density and distribution of the ions and excited molecules about the path of each incident particle. The incremental energy loss with distance is called the *linear energy transfer* (LET) and is a function of the velocity v and numerical charge z of the ionized particle according to the Bathe formula [2]:

$$-dE/dx = (4\pi e^4 z^2 NZ/mv^2) \ln (2mv^2/L) \tag{2.1}$$

where NZ refers to the number of electrons per unit volume of the adsorber, e and m are the charge and the mass of the electron, and L is the mean excitation potential of the adsorber, which for all practical purposes can be set equal to its ionization potential. This relation is valid for electrons below relativistic energies, i.e., below $mc^2/2$ or 0.255 MeV. For an electron of energy E (in electronvolts) traveling through polyethylene, the LET is given in units of electronvolts per angstrom by the numerical relation $(980/E) \log 0.2E$. Thus the LET is about 0.01 eV/Å when $E = 0.25$ MeV, and it increases to 2.3 eV/Å at $E = 1$ keV [3].

The probability of ionization and excitation depends on the velocity of the electron, and it increases rapidly as the electron slows down toward the end of its path. However, the energy lost during such a collision is greatest early on in the path, when the primary electron has its maximum energy. A consequence of the increase in LET with decreasing energy is that the spacings between consecutive excitations along the main track decrease from a few thousand angstroms at an electron energy of 1 MeV to the order of 100 Å at 5 keV. Below 1 keV, the successive excitation processes that are produced in the wake of the electron's path begin to merge together, so that in principal terms the distribution changes from a ''string of beads'' above 1 keV to a continuous ribbon at lower energy [3]. As long as the kinetic energy of the electron is well above the ionization potential (10–15 eV for organic molecules), there is always a high probability that molecular ionization will occur in each of these excitation events, resulting in the production of secondary electrons. These secondaries are called δ-rays when they have sufficient energy to produce further secondaries. Owing to their low velocity, δ-rays have short tracks that create local regions of high-ionization density which eventually give rise to high-radical density also. The δ-rays dissipate their energy in the same manner as the faster electrons, except that the clusters of ions they produce are closer together because δ-rays are less energetic than primary electrons. The integrated track length of all the δ electrons is only about 2.5% of that of a primary electron between 10 and 400 keV, and their average energy is small, only about 5% of the ejected electrons having

energies in excess of 100 eV and 1% in excess of 500 eV. However, δ-rays cause
about half the total ionization from fast electrons because their number is considerably greater than the number of electrons in the primary beam and some of them carry
essential energy and produce further low-energy secondaries along their own tracks.
After the kinetic energy of a secondary electron has fallen below the ionization potential of the medium, the electron will be incapable of causing further ionization,
but instead will cause electronic excitation. After this excitation, the energy of the
electron will have fallen to a point where it is below the lowest level of electronic
excitation of the medium. Such an electron is known as a *subexcitation electron,* and
its energy is dissipated during the process of the excitation of molecular vibrations
until its kinetic energy drops below that of the lowest vibrational quantum. If, as is
often the case, the lowest vibrational quantum is on the order of 400 cm^{-1}, the
electron will have an energy of less than 0.05 eV and will be known as an *epithermic
electron.* The difference between this energy and kT (0.006 eV at 77 K) will be
dissipated by molecular vibration in the condensed phase. This is the last stage of
electron *thermolization.* During energy loss, the electron will be displaced, and at
that time it becomes thermic, that is, when the electron has a kinetic energy equal to
kT, it will be at a certain distance r_t from the ion from which it has been ejected
(parent ion) [4].

The processes described lead to the formation of compact groups of ions and excited molecules commonly referred to as *spurs,* with the distances of positive ion–
electron separation r_t in the condensed state in the range 50–100 Å. Mozumder and
Magee [5] classify the spatial distribution of sequential events, including both ionization and excitation, into three regions comprized of isolated spurs, short tracks,
and blobs, according to the mean separation distance between the individual spurs.
Isolated spurs are considered as noninteracting entities separated in space. *Short tracks*
are defind as regions in which consecutively formed spurs begin to overlap. In the
blob, the spurs merge together to form a larger continuous region of high-ionization
density. The fractions of the total energy dissipated in isolated spurs, short tracks,
and blobs are computed to be 0.6, 0.25, and 0.15, respectively [6].

Unlike the situation in an irradiated gas, the electron liberated by ionization of a
molecule in the condensed phase may have insufficient kinetic energy to escape the
field of its parent ion. The critical escape distance r_c beyond which this electron can
be considered free, i.e., outside the influence of its parent ion, can be estimated [7]:

$$e^2/\epsilon r_c = \tfrac{3}{2}kT \tag{2.2}$$

where ϵ is effective dielectric constant. In organics, with a bulk dielectric of about
2, the critical distance becomes 185 Å at 300 K. One can therefore expect that only
a small fraction of the ejected electrons can truly escape. If the irradiation is carried
out at relatively high temperatures, the secondary electrons will, after having spent
their kinetic energy, be drawn back to and recombine with the parent ion or any
other positive ions in the vicinity to give highly excited molecules.

2.2 Transient Intermediates in Chemical Reactions Induced by High-Energy Radiation

As has already been mentioned, radiations transfer their energy to matter through electostatic interaction of fast-moving electrons with the orbital electrons of the irradiated substance, and as a result, a nonuniform distribution of intermediates is laid down by the radiation. (Specific effects which arise when neutrons are involved will not be considered here.) Some of the species will react in the spur, whereas the remainder will undergo diffusive separation to achieve a uniform spatial distribution and so contribute to the steady-state population. If the energy transferred to the orbital electron is less than that required for the ionization process, the electron is raised to an upper level, giving rise to an excited state:

$$AB \rightarrow AB* \qquad (2.3)$$

Other excited molecules arise as a result of charge neutralization:

$$AB^+ + e^- \rightarrow AB* \qquad (2.4)$$

If, however, the energy transferred by the incident electron to a particular orbital electron is higher than its binding energy, this electron may be expelled, leaving behind a positively charged ion. This process of ionization can be schematically written

$$AB \rightarrow AB^+ + e^- \qquad (2.5)$$

Ionization may lead to molecular fragmentation, as is well known from mass spectroscopy:

$$AB \rightarrow A^+ + B + e^- \qquad (2.6)$$

In Eq. (2.6), B is a neutral fragment which may be a free radical.

It thus appears that the active species resulting from primary interaction of radiation with matter include excited molecules, positive ions, electrons, and free radicals. All radiation-induced reactions result from the interaction of these active species with themselves or with the molecules of the surrounding medium.

2.2.1 Behavior of the Excited Molecule

In an excited molecule, the energy is increased by the higher energy level of one or more electrons. High-energy radiation produces a broad range of excitation energies and, consequently, a large number of possible reactions.

Atoms have only electronically excited states to be considered. Polyatomic molecules, however, show electronic, vibrational, and rotational excitations, and every electronically excited state has an accompanying set of vibrational and rotational states. On an energy scale, differences between rotational states are represented by about 0.01 eV, vibrational states by about 0.1 eV, and electronic states by about 3 eV. In absorption spectra, rotational excitation shows as absorption in the microwave and far-infrared region, vibrational excitations absorb in the near-infrared region, and electronic excitations absorb in the ultraviolet region [8]. The excitation is of the whole molecule, not just one part of it, and in large molecules, reaction can occur at a site remote from the track of the particle. The same phenomenon occurs in photochemistry, where energy is absorbed by the chromophore and yet reaction or emission of fluorescence can occur elsewhere in the molecule.

   The processes by which a molecule in an excited state can dissipate its energy are governed by the Franck-Condon principle [9]. This principle is based on the great difference in mass between an electron and a nucleus. When an electron transition occurs, either as a result of absorption of a photon or by interaction with a charged particle of high energy, the time taken is very much smaller than is needed for any appreciable change in the distance between nuclei. For diatomic molecules, the electronic and vibrational states can be represented on a *potential-energy diagram*, a curve representing the potential energy of the molecule as a function of the interatomic distance. More complicated molecules cannot be represented directly by the two-dimensional figures. However, as a very rough approximation, such a figure may be considered as representing the potential energy of one bond between two groups in a polyatomic molecule. In Fig. 2.1, the energy levels for the ground and excited

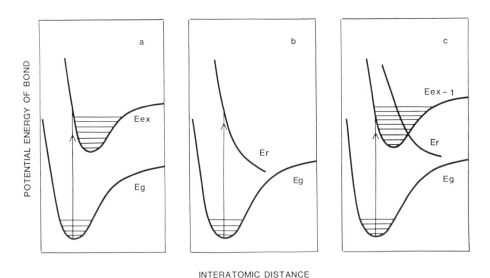

INTERATOMIC DISTANCE

Fig. 2.1. Schematic representation of dissociation *(a,b)* and predissociation *(c)* processes for a diatomic molecule. *(a)* Excitation to a point in the electronic state *(Eex)* above the dissociation limit. *(b)* Excitation to an electronic state *(Er)* which is repulsive, i.e., has no minimum in potential energy and is dissociative at all points. *(c)* Excitation to a vibrational level in the electronic excited state (Eex − 1) which is higher than the crossing point of a repulsive state *(Er)*. After [8], by permission of Academic Press.

states of a diatomic molecule are shown in terms of the distance between nuclei. The molecule has an equilibrium interatomic distance at the minimum in the curves. The horizontal lines on the curves represent a few of the vibrational levels in each of the electronic states. On excitation, a molecule initially in a ground state will *dissociate* if it is excited to a vibrational or rotational level above the dissociation limit for the state *(a)* or to a repulsive state (no minimum in the potential-energy curve) *(b)*. Under these conditions, dissociation may take place extremely rapidly—in a time on the order of one atomic vibration, or 10^{-13} sec [10].

A second possibility, termed *predissociation*, may arise when the excited level would itself give a stable molecule, but it intersects a repulsive state which allows dissociation *(c)*. The molecule in vibration will at some point in the vibrational cycle be at the same interatomic distance as that of the dissociative state, and sooner or later the crossover will occur. Because the vibrational levels do not match exactly, the crossing over from one state to the other may be as slow as 10^{-9} sec.

Both dissociation and predissociation result in decomposition of the molecule into radicals each with excess kinetic energy:

$$AB^* \rightarrow A^. + B^.$$ (2.7)

When polyatomic molecules are considered, the situation becomes far more complex because of the great variety of possible levels for electronic, vibrational, and rotational energy. Owing to the large number of degrees of freedom, a considerable time may then elapse before conditions are suitable for decomposition to take place. This will increase the probability of changes in the subsequent distribution of energy between these states by intersection of the corresponding energy levels, and the absorbed energy may be reallocated over a number of vibrational states. The process when electronic energy is converted to vibrational energy is called *internal conversion*. The vibrational energy may then be dissipated among a number of degrees of freedom, giving the equivalent of a molecule at high temperature. In the other words, internal conversion degrades the electronic energy into heat:

$$AB^* \rightarrow AB + heat$$ (2.8)

Deactivation of the excited state also may occur by the emission of light:

$$AB^* \rightarrow AB + h\nu$$ (2.9)

As was discussed in Chapter 1, it can be either fluorescence or phosphorescence. In polyatomic molecules, however, deactivation of vibrational levels by radiationless transitions is quite common because the vibrational energy in one bond can be dissipated in the other bonds of the molecule.

2.2.2 Fate of the Free Electron

In low-dielectric condensed media, most ions that are initially produced by radiation do not survive long but undergo neutralization. However, a small, yet significant number always escape depending on temperature, dielectric constant, radiation quality, etc. In radiation chemistry, radiochemical yield is expressed with respect to an absorbed energy dose of 100 eV by the symbol G. The G value is by definition the number of molecular changes occurring per 100 eV absorbed in the system. It is believed that G for total ionization is about 40 times higher than that of escaped ions [15]. No simple technique exists to discover the eventual fate of the escaped electron, but it is suggested that it, after losing most of its excess energy in further excitation or ionization or by collision, is trapped by another positive ion or by a molecule of high electron affinity such as oxygen or a halide. Along with this, radicals presented in solids are very efficient deep traps for electrons:

$$R^{\cdot} + e^{-} \rightarrow R^{\cdot -} \tag{2.10}$$

2.2.3   The Nature of Positive Ions

Ion molecules can be produced either directly, by stripping off an electron, or indirectly, from a highly excited molecule. The *ionization potential* for an individual atom is the minimum energy required to detach an electron, and it ranges from 3.9 eV for cesium to 24.6 eV for helium. Multiple ionization requires higher energies. With high-energy radiation, though, this is not necessarily a very rare occurrence, and doubly ionized atoms may appear at about 1–10% of the frequency of singly ionized atoms [8]. With complicated molecules, however, such doubly charged ions in most cases undergo immediate fragmentation.

An excited molecule may have more energy than is required to cause ionization if the electron involved has been removed from an inner level. The excess energy may be lost in several ways, e.g., directly by emission of fluorescence, indirectly by collision, or by internal transference to another electron energy level to give rise eventually to an ion. The situation is somewhat similar to that described above for excited states leading to predissociation, and it is termed *preionization*. In an excited polyatomic molecule, the time elapsing before the molecular ion is formed allows for some selectivity in the course of the reaction. Although any electron may be excited to a high level by high-energy radiation, certain forms of ionized structure may therefore be favored.

Ions are usually formed with considerable excess of vibrational energy; they may dissociate either in a single thermal vibration (if simple in structure) or over longer periods in more complex systems where the vibrational energy is distributed over a number of degrees of freedom.

In the general case, the initial ionization leaves either an excited- or a ground-state molecular ion. If an excited state is formed, it can dissociate before recombination:

$$M^{+*} \rightarrow R_1^{\cdot +} + R_2^{\cdot} \tag{2.11}$$

where $R_1^{\cdot}$ and $R_2^{\cdot}$ are a radical ion and a radical, respectively. The dissociation also can proceed by molecular disproportionation to yield a molecule ion and molecule:

$$M^{+*} \rightarrow M_1^{+} + M_2 \tag{2.12}$$

One of the molecules, M_1 or M_2, will be unsaturated, and the gain in energy from the formation of the double bond will make the process exothermic. In both these processes, the charge is generally carried by the fragment of lower ionization potential. The molecular dissociation requires an atom transfer (usually hydrogen); therefore, Reaction (2.12) is slower process than Reaction (2.11), which can occur in a single vibrational cycle. With many organic compounds, however, molecular fragmentation predominates and lower-molecular-weight molecule ions result.

In a system containing two types of molecules, charge exchange can occur with the molecule of lower ionization potential:

$$M^+ + A \rightarrow A^+ + M \tag{2.13}$$

Such an exchange in the liquid state demands a long life for the positive ions. Resonance charge exchange with like molecules is also a mechanism by which ions can be rapidly diffused away from the spurs.

The positive ions formed in the track and spurs must eventually be neutralized. The time for this process in the liquid and solid states has been estimated to be of the order of one vibrational cycle (10^{-13} sec) or somewhat longer [11]. In the gas phase there is no question that a much longer time cycle applies, since the electron escapes from the vicinity of the positive ion immediately. The main difference in the effect of short or long time cycles is whether or not ion dissociations occur before neutralization and whether or not charge exchange with molecules of lower ionization potential can occur before neutralization. Owing to fast neutralization, ion reactions are probably not the predominant feature of the radiolysis mechanism in most liquid and solid systems.

2.2.4 Radicals Produced by Irradiation

Radicals are defined as such molecules or fragments in which the chemical bond is modified to result in an electron having uncompensated spin. The position in the molecule where this modification occurs is denoted in chemical formulas by a dot. The symbol used in structural formulas does not mean that the unpaired electron is located at this place; it means only that its probability density is usually highest there. Electron-spin resonance data show that the unpaired electron in radicals is delocalized over a large part of the molecule and is in contact interaction with many nuclei in the molecule. In aliphatic radicals, this delocalization extends over 6–8 C—C bonds; in conjugate systems, it usually extends over the whole molecule; and in solids, the delocalization may reach even macroscopic measures.

An excited molecule can dissociate into two radicals or into two molecules:

$$M^* \rightarrow R_1^{\cdot} + R_2^{\cdot} \tag{2.14}$$
$$M^* \rightarrow M_1 + M_2 \tag{2.15}$$

Any excess energy can be partially converted into kinetic energy. Radicals formed with high kinetic energy are termed *hot radicals* and may undergo reactions which would not normally take place because of high energy of activation. When molecular dissociation (2.15) occurs, the molecules represent final products of the radiolysis. If radical dissociation occurs, the free radicals formed may engage in the following types of reactions:

1. *Radical-radical recombination reactions.* These occur primarily in the spurs, where the concentration of radicals is high:

$$R_1 + R_2 \rightarrow M \tag{2.16}$$
$$R_x + R_y \rightarrow M_3 \tag{2.17}$$

Reaction (2.16) represents a recombination of radicals to the original molecule and results in no net effect. Reaction (2.17) represents the reaction of two radicals that combine to form a new molecule, different from M. The activation energy for such a radical-radical combination is generally quite low and may be zero for many radicals, but there are static factors which may reduce the probability of recombination.

2. *Radical-radical disproportionation reactions.* These usually occur with transfer of a hydrogen atom. Thus two ethyl radicals may combine to form butane by Reaction (2.17) or disproportionate to form ethylene and ethane by Reaction (2.18):

$$C_2H_5 + C_2H_5 \rightarrow C_2H_4 + C_2H_6 \tag{2.18}$$

Such reactions have small activation energies and are highly exothermic because of the energy gained in forming the double bond. The two radicals need not be identical for disproportionation reactions to take place.

3. *Abstraction reactions.* Radicals in the spur or radicals that escape from the spur may react with molecules to abstract a hydrogen atom. Such reactions must be exothermic to occur. They have activation energies of 5 to 10 kcal mol^{-1}

$$R + MH \rightarrow M + RH \tag{2.19}$$

The radical M cannot react with molecules MH by the mechanism in Reaction (2.19) because this would produce no net change. Therefore, M radicals have a long life and disappear only by radical-radical combinations, radical dissociation, or by reaction with radical scavengers. M radicals can disappear by recombination or by addition to the double bond in unsaturated compounds. If M is a large radical, a decomposition into a small molecule and a radical can occur:

$$M \rightarrow M_1 + R_1 \tag{2.20}$$

The activation energy of this process is high for hydrocarbon radicals (20 to 40 kcal mol^{-1}), but reaction occurs with a low activation energy for large polyhalogen radicals, in which case R_1 is usually the halide atom.

Certain molecules possess a high affinity for radicals and may react with them on almost every collision. For example, the oxygen molecule, which has two unpaired electrons (a diradical), reacts readily with free radicals:

$$R + O_2 \rightarrow RO_2 \tag{2.21}$$

Thus irradiations in which oxygen is present yield less of some products but may yield more of others, the yield depending on how the products of Reaction (2.21) enter into the overall reaction sequence.

Free radicals are extremely important in radiation chemistry. There is considerable evidence that in many organic systems the G values for radical production (resulting from excitation) run to about 3–5, whereas the G values for ion formation are down by a factor of 10 or more [12]. Most of the analyses of radiation-induced changes assume that although the chemical nature of the radicals produced is fairly specific, they are located at random throughout the specimen. This considerable simplification is theoretically justifiable only in certain cases. Along the track of a fast electron, ionization and excitation will occur in small clusters, associated with the production of δ electrons, and this will give rise to radicals in close proximity. Furthermore, radicals formed along the same primary track are produced simultaneously, whereas the overlap of separate radicals from two primary tracks will usually occur at much longer time intervals (depending on the radiation intensity). Reactions between radicals formed in the same track will therefore be favored, and this will itself tend to produce effects which are largely independent of radiation intensity.

Each primary track may be considered as a cylinder within which the radicals have diffused at random. If during the radical lifetime τ a number of primary tracks have overlapped, interactions between radicals from different tracks are highly probable, and a random distribution is a reasonable description. This situation is favored by high-intensity radiation and (for a given intensity) by sparsely ionizing radiation, such as fast electrons or γ-rays. Conversely, if the radical lifetime τ is small compared with the time between the passage of successive primary tracks through the same volume element (determined by the effective diameter of the track), radical reactions will be confined primarily to each separate track. Experimentally, however, it is usually found satisfactory to assume a random distribution of the radicals.

2.3 Energy Transfer

The process by which the energy captured at random throughout a system is directed into those chemical bonds or, in the case of mixtures, into those compounds which are most susceptible to radiation is called *energy transfer*. The term *energy transfer* covers a wide variety of processes, including any process whereby one molecule which has been affected by radiation produces an effect in another molecule. On the basis of this definition, charge transfers or free-radical reactions would count as energy transfers even though the processes do not necessarily involve the transfer of energy. From this point of view, a more suitable expression might be *reactivity transfer* [1].

Transfer of energy between an excited molecule (or a molecular ion) and its neighbors can take place in alternative ways. Thus an excited molecule A* can transfer its energy to a neutral molecule or atom B, causing excitation in the latter, provided that the excited state of B requires the same or less excitation energy than does A:

$$A* + B \rightarrow A + B* \tag{2.22}$$

If not quickly deactivated or converted to vibrational energy, B* may radiate its own characteristic fluorescent spectrum even though the energy was initially absorbed by A. The excitation energy also can be shared between A and B, giving a lower excitation level to A:

$$A* + B \rightarrow A^{1}* + B* \tag{2.23}$$

Or it may be sufficient to ionize B, if the ionization energy of B is sufficiently low:

$$A* + B \rightarrow A + B^{+} + e^{-} \tag{2.24}$$

Further reactions include such chemical changes [13] as:

$$A* + BC \rightarrow AB + C* \tag{2.25}$$
$$A* + B \rightarrow AB \tag{2.26}$$

In the last case, a third body must be present to carry off excess energy. Reactions of this character might account for the incorporation of an additive into the structure of a polymer molecule.

In a perfect crystal of individual molecules, energy transfer may occur by a process termed *excitation migration*. The resonance between adjacent molecules in a sense delocalizes the excitation energy, which may be transferred over long distances to react with the small amount of additive present. In polymers, energy transfer occurs predominantly along the molecular chains, over distances of ~ 1500 Å [14].

In an ionized molecule, the electron vacancy need not be considered as localized, but rather as spread throughout the molecule, with a varying probability depending on its structure. In such an ionized structure, the bond valency rules obeyed by the neutral structure need not apply, and changes in configuration may readily occur. To some extent, the same considerations may apply to a radical molecule which, although neutral, has an extra energy level available, permitting some electron mobility. The concept of a mobile electron vacancy gives the possible explanation for energy transfer within an ionized molecule. Energy (or charge) transfer between molecules of the type

$$A^{+} + B \rightarrow A + B^{+} \tag{2.27}$$

can occur with high efficiency if the reaction is exothermic. Depending on the potential curves of A and B, it may be associated with the emission of radiation or with dissociation due to the change in positions of the nuclei in the ion and the molecule.

Except in special systems, charge transfer of an electron

$$A^{-} + B \rightarrow A + B^{-} \tag{2.28}$$

is far less likely, since the capture of a free electron is a very selective process and will usually occur by the molecule with the highest electron affinity.

References

1. Charlesby, A.: Atomic Radiation and Polymers. Pergamon Press, New York, 1960
2. Fano, U.: Ann. Rev. Nucl. Sci. *13*, 1 (1963)
3. Williams, F.: The Radiation Chemistry of Macromolecules, vol. 1 (ed. Dole, M.). Academic Press, New York, 1972
4. Magat, M.: J. Chimie Phys. *63*, 142 (1966)
5. Mozumder, A., Magee, J.L.: Radiat. Res. *28*, 203 (1966)
6. Mozumder, A.: Advances in Radiation Chemistry, vol. 1 (ed. Burton, M., Magee, J.L.). Wiley, New York, 1969
7. Veisberg, S.E.: Radiation Chemistry of Polymers (ed. Kargin, V.A.). Nauka, Moscow, 1973
8. Newton, A.S.: Radiation Effects on Organic Materials (ed. Bolt, O., Carroll, J.G.). Academic Press, New York, 1963
9. Condon, E.V.: Am. J. Phys. *15*, 365 (1947)
10. Swallow, A.J.: Radiation Chemistry of Organic Compounds. Pergamon Press, New York, 1960
11. Samuel, A.H., Magee, J.L.: J. Chem. Phys. *21*, 1080 (1953)
12. Freeman, G.R., Fayad, R.J.M.: J. Chem. Phys. *43*, 86 (1965)
13. Franck, J., Platzman, R.: High Energy Radiation, part 1. McGraw-Hill, New York, 1954
14. Partridge, R.H.: J. Chem. Phys. *52*, 2491 (1970)
15. Blake, A.E., Charlesby, A., Randle, K.J.: J. Phys. [D] *7*, 759 (1974)

Chapter 3

Thermoluminescence in Polymers Induced by Radiation

Ionizing radiation, as it passes through matter, excites and ionizes the molecules surrounding its trajectory. The secondary electrons which are extracted from the molecules during ionization progressively transfer their kinetic energy to the medium until they become thermic. A thermic electron will be at a distance r_t from the ion from which it has been issued (parent ion). Several cases may be cited [1]:

1. If the distance r_t is less than a certain critical distance r_c to which the Coulomb attraction is exerted by the parent ion (that is, $r_t < r_c$), then the electron will be drawn toward the parent ion, which it will neutralize and form into a highly excited molecule. Such highly excited molecules formed by neutralization eventually become inactive, whether it be by decomposition, internal conversion, collision, or a combination of two or more of these factors. The return to the parent ion is very rapid. The time which elapses between the process of ionization and the process of neutralization can be estimated at less than 10^{-10} sec. If the neutralization is followed by an emission of light, this emission will occur 10^{-7}–10^{-8} sec after ionization if the emission takes place beginning with a singlet state and 10^{-3} sec to several minutes if it is from a triplet state.

2. If $r_t > r_c$, the electron breaks free and may recombine not necessarily with its parent ion, but with any of the ions present in the medium. Under normal conditions, the time between ionization and neutralization may be estimated at 10^{-2}–10^{-3} sec. In the two cases which we have just examined, light emission is produced during irradiation. Such an emission is called *radioluminescence*.

3. During and particularly at the end of the thermolization process, the electron may become "trapped," and its recombination with an ion (and, consequently, the emission of light) may be slowed down. This retardation is highly variable. In the case of solid media, it may be measured in minutes, hours, or sometimes even days. Regardless of whether or not the secondary electron is temporarily trapped, it will in time recombine with a positive ion and charge neutrality will again be established.

This event, as well as the reactions induced by the electron during its travel through the matrix, will lead to a proliferation of stable and unstable products. Some of the electronically excited molecules produced in this manner will return to their ground state by luminescence emission and by inter- or intramolecular energy transfer. The photons given off by the sample can thus come from the excited molecule or from chemicals attached to or contained in the polymer. It is this process, described as *deferred radioluminescence,* with which we are presently concerned and which we will be examining in more detail.

The most immediate electronic processes of ionization, excitation, and electron capture are not influenced to any significant extent by the viscosity of the medium. Turning now to the processes which follow excitation and ionization, those involving molecular dissociation and reaction without mass transfer do not seem to be greatly influenced either by the rigidity of the medium or by molecular size [2].

However, all processes which depend on atomic (hydrogen atom) or molecular diffusion are profoundly affected by the viscosity of the matrix. After charge separation has been achieved, the subsequent reactions of stabilized electrons appear to be governed largely by molecular or segmental diffusion. The practical importance of molecular motion is illustrated by the considerable variation in the lifetimes of stabilized electrons according to the viscosity of the medium. The experimental results suggest an interpretation of the thermal decay of trapped electrons in terms of a model consisting of various populations of trapped electrons that are relatively stable toward decay until the temperature is raised to a point where they become vulnerable [3]. The activation energy of this process in polyethylene, for example, is 3–5 kcal mol^{-1} in the 77–127 K temperature interval and is not inconsistent with some type of rotation about a C—C bond. Hence it is feasible that the thermal decay results from some kind of molecular motion involving the surrounding environment of the trapped electron which causes the electron to become mobile. The frequency factor for the decay of the trapped electrons in polyethylene has a value of 10^5 sec^{-1}. Such a value is reasonable for the complicated motion of a polymer chain. Bond vibrations correspond to frequency of the order of 10^{13} sec^{-1} and rotations to 10^{11} sec^{-1}. Therefore, a more complicated motion, perhaps of a torsional type, would be expected to take place with a lower frequency.

If irradiation of the polymer is carried out at low temperature (say, at liquid nitrogen temperature), the reactive species are frozen in the solid and their concentration builds up as the irradiation proceeds. The release of electrons from their traps is temperature-dependent. On heating the sample, an erosion of the traps is induced by the onset of local motion in the polymer involving molecules located in the immediate vicinity of the electron trap. The recombination of ions on heating of an irradiated sample produces the electronically excited molecule M*:

$$h^+ + e^- \rightarrow M* \tag{3.1}$$

In organic substances, this reaction is determined not by the thermal liberation of captured electrons, but by the migration of electrons and positive holes together with their stabilization centers at the moment of "unfreezing" the mobility of the medium.

The excited molecule generated in the process of charge recombination can return to the ground state by means of light emission:

$$M^* \rightarrow M + h\nu \qquad \text{(fluorescence: } S_1 \rightarrow S_0, \text{ or phosphorescence: } T_1 \rightarrow S_0) \qquad (3.2)$$

These schemes have been verified by observing the sensitivity of the thermoluminescence to optical illumination at low temperatures, referred to as *photobleaching* [4], and by observing the stimulating effects of an electric field [5].

3.1 Thermoluminescence Induced by Ionizing and Ultraviolet Radiation: The Notion of a Glow Curve

Luminescence observed as a result of the irradiation of a substance at a low temperature and its subsequent warmup is called *radiothermoluminescence,* and the resulting plot of emission intensity against temperature (or warming time) is called a *glow curve*. Thermoluminescence can be induced either by high-energy or ultraviolet (UV) radiation. The term *radiothermoluminescence,* strictly speaking, presumes the excitation by ionizing radiations, whereas the excitation by UV radiation should be referred to as *photothermoluminescence.*

There are some basic differences between the excitation by high-energy and UV radiation:

1.   When excitation is achieved by UV radiation, the absorption of a single photon results in the primary excitation of only a single molecule. UV radiation is absorbed by individual chromophores, e.g., those containing double bonds and carbonyl groups. It also may be selectively absorbed by aromatic impurities. However, ionizing radiation produces excitation in many molecules before its total energy is absorbed. Thus high-energy radiation interacts with the whole matrix in a statistically equivalent manner, and there is no selectivity of absorption peculiar to one particular type of molecule.

2.   UV radiation produces the excited species randomly throughout the volume. High-energy radiation generates the species in clusters along the path of the radiation.

3.   Light may excite a molecule to only one of a few of its possible excited states, whereas passage of ionizing radiation can, in principle, produce all possible excited states of the molecule and the molecule ion.

Thus the initial situation resulting from the passage of an ionizing particle is much more complicated than that resulting from the absorption of light.

High-energy and UV irradiations of polymers generally result in different glow curves. This is characteristic not only for polymers, but for quartz [6] and ionic crystals [7] as well. In the latter case, it was shown that both X-rays and UV radiation populate the same sites but with different degrees of probability.

In many cases, the radiant energy of the UV radiation is insufficient to cause direct ionization, and ionization occurs by a double-excitation process [8]. In this process, a molecule is raised to the triplet state, via intersystem crossing from the lowest excited singlet state by a first photon, and then it has an appreciable probability of absorbing a second photon while still in the triplet state (due to its long life-time, often several seconds), which would supply enough extra energy to cause complete ionization. This process will occur particularly in molecules with a high singlet-triplet crossover efficiency.

Simultaneously with producing excitation and ionization, UV light leads to the emptying of some of these traps, giving rise to what is known as *bleaching*. Although bleaching can be expected to also take place in the case of high-energy radiation, the untrapping rate in the case of UV radiation is about 10 times higher, because the UV/visible radiation is directly absorbed by the trapped electrons, whereas the ionizing radiation probably has to interact indirectly through degradation of the incident energy into local thermal motion of polymer chain segments which then "shakes free" the electrons from their shallow traps.

The processes described lead to redistribution of the thermoluminescence intensities along the temperature scale. Figure 3.1 shows glow curves for highly crystalline poly-3-methylbutene-1 after exposure at 77 K to UV and X-ray radiations as well as consecutive exposure to X-rays and then, UV radiation [9]. It can be seen that exposure to UV radiation as well as consecutive exposure to two types of radiation completely "bleaches out" the low-temperature thermoluminescence peaks at 115 and 125 K. A similar effect was observed by Charlesby and Partridge [10] for a variety of polyethylene samples. They noticed that exposure of the samples to UV radiation gives much weaker low-temperature luminescence (77–127 K) than exposure to γ-radiation.

3.2 Evidences for the Ionic Nature of Radiothermoluminescence in Organic Substances

In general, it seems that there can be three main causes of radiothermoluminescence in solid materials. These are molecular excitations, chemical reactions, and charge trapping followed by recombination [11].

Molecular excitation has been found to be a very common cause of organic luminescence. The greatest objection to this as a cause of radiothermoluminescence is the comparatively short decay half-lives involved. The maximum half-lives observed in a wide variety of organic materials lasted a few seconds [12], whereas the half-life of radiation-induced luminescence at liquid nitrogen temperature is many orders of magnitude larger (samples usually give considerable emission even if kept for many days between irradiation and warming).

The most obvious chemical reaction mechanism would be radical-radical recombination. This can be ruled out as the cause of luminescence on a number of grounds [13]. The variation in thermoluminescence intensity with dose usually reaches a maximum at a few megarads, but radical production is linear with dose up to more than 50 Mrad [14]. No emission is observed when radicals are produced at low temperatures by purely mechanical means (grinding) and the sample is then warmed up. The thermoluminescence output can usually be reduced to nearly zero by illuminating the sample with visible light prior to warming (optical bleaching), whereas this has no

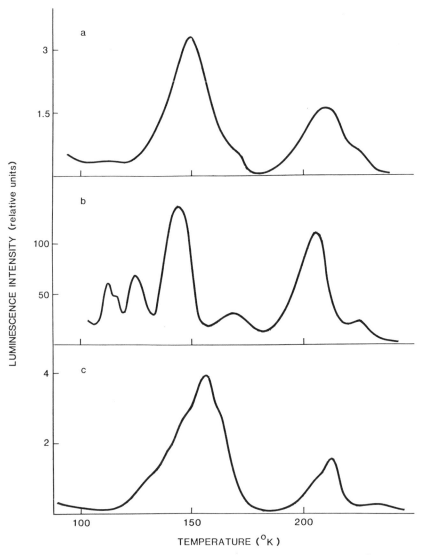

Fig. 3.1. Glow curves for highly crystalline poly-3-methylbutene-1: (*a*) after exposure to UV radiation; (*b*) after exposure to X-rays; and (*c*) after consecutive exposure to X-rays and then to UV radiation. After [9], by permission of Plenum Publishing Corp.

effect on radical concentration. The electron-spin resonance investigation of alkyl radical concentration in relation to thermoluminescence emission showed that there is no increase in the rate of disappearance of alkyl radicals in the temperature neighborhood of the thermoluminescence peaks and that alkyl radicals continue to decay long after the luminescence is exhausted [13]. The kinetic of radical–radical recombination must be of the second order, whereas radiothermoluminescence has been shown to follow first-order kinetics. In addition, free radicals are generally completely "frozen in" at liquid nitrogen temperature, so they cannot react with each other. Moreover, the very small activation energy of the thermoluminescence process at temperatures not essentially higher than liquid nitrogen temperature makes any type of chemical reaction unlikely as the source of the thermoluminescence.

The only remaining possible cause of radiothermoluminescence in organic solids is the recombination of trapped ions. As will be shown, all phenomena accompanying radiothermoluminescence are explicable if trapped ions are the reactive species, and there is no doubt that thermoluminescence in irradiated organic substances is caused by ion recombination.

3.3 Factors Determining Radiothermoluminescence

3.3.1 The Nature of Electron Traps

Macromolecular systems are usually characterized by a high viscosity, especially in the solid state at low temperatures. Consequently, there is a large probability that intermediate species generated during the irradiation of polymers remain trapped in the host material, their reaction rates being essentially limited by the slow diffusion process that applies in such circumstances. In particular, stabilized electrons are generally produced in condensed systems by the polarization of surrounding molecules.

Traps in polymers can be of different natures, either physical or chemical. An electron physically trapped in a polymer matrix can be considered as a distinct entity by itself; i.e., it is not localized in a specific molecular orbital belonging to a single molecule or its segment. Such an electron is localized in cavities or voids which are associated with structural imperfections in disordered solids and is only weakly coupled to its environment. The concentration and distribution of physical traps depend to a great extent on the sample prehistory. One should expect a continuous depth distribution, because of various degrees of perfection for the same type of a trap, as well as exponential (or pseudoexponential) distribution, because large structural defects are formed with an essentially lower probability than small ones.

Differently from a physically trapped electron, a chemically trapped electron can be regarded as being bound to some particular molecule and to reside entirely within the characteristic molecular orbitals of that molecule. A chemically trapped electron would therefore be expected to undergo significant interaction with the protons (or other magnetic nuclei) of its host molecule. Chemical traps are characterized by a discrete depth distribution, reflecting the electron affinity of the various atomic groups representing chemical traps. A second peculiarity of traps of a chemical nature is the independence of their concentration and distribution of the conditions of sample preparation (quenching, annealing, the rate of temperature change, etc.), although free radicals and some volatile impurities (dissolved gases) are the exception.

Electron-spin resonance and optical spectroscopy techniques provide the most reliable methods for identifying trapped electrons in polymers. Optical spectra of γ-irradiated hydrocarbon samples obtained at 77 K before and after illumination with light of different wavelengths indicated that both physically and chemically trapped electrons can be photobleached. For chemically trapped electrons (those formed as a result of electron capture by aromatic solutes, for example) this effect normally requires photon energies corresponding to the visible and UV regions, whereas physically trapped electrons can be photobleached by light in the near-infrared region [3].

The linewidth of the electron-spin resonance singlet spectrum from the trapped electron was found to be much narrower (especially for physically trapped electrons) than the linewidth associated with the spectra of most organic radicals in the glassy state. This means that the narrow singlet is rather easily discriminated from the broad underlying spectra of other paramagnetic species, even when the latter are present in much higher concentrations [15]. As a result, yield measurements and kinetic studies on trapped electrons can be carried out by electron-spin resonance spectroscopy with high accuracy. As with optical measurements, complete removal of the electron-spin resonance singlet can be achieved by photobleaching with suitable light, paralleling the observations made by optical spectroscopy.

To date, the most comprehensive study of the nature of trapping sites has been carried out for polyethylenes [16]. Measurements of trapped electron concentrations on polyethylene specimens γ-irradiated at 77 K showed that trapped electron concentration increases to a maximum at a dose of about 0.5–1 Mrad and then decreases steadily. There is very little thermal decay of trapped electrons at 77 K, so the electrons must disappear by some radiation-induced mechanism. By contrast, the free-radical concentration increases progressively with dose. This indicates that electrons disappear from the system by reaction with the free radicals which build up during irradiation. The possibility that some other radiation products whose yield is proportional to dose might be responsible for the effect can be ruled out, since annealing the sample to the point where virtually all the free radicals have been destroyed restores the electron-trapping ability of the matrix.

The way in which free radicals may influence electron trapping is by the formation of a carbanion $R^{\cdot-}$:

$$R^{\cdot} + e^{-} \rightarrow R^{\cdot-} \tag{3.3}$$

If it is assumed that the product of this reaction, the carbanion $R^{\cdot-}$, occupies the trapping site, then further electron trapping at this site would be precluded. Since $R^{\cdot-}$ is diamagnetic, the trapped electron concentration, as observed by electron-spin resonance, would be expected to decrease when the free-radical concentration has reached a level where reaction can occur to a significant extent. It should be pointed out that reactions of the type represented by Eq. (3.3) can be expected to be exothermic by about 1 eV or more [17], as judged by analogy with the electron affinity of the methyl free radical, and should therefore be thermodynamically feasible. Additional experimental support for the existence of carbanions in irradiated hydrocarbons has been obtained by Ekstrom *et al.* [18], and it was suggested that the reaction of the radical ions $R^{\cdot-}$ and molecular ions M^{+} rather than electron-hole recombination is responsible for the formation of excited molecules and following luminescence [19,35].

Optical spectroscopy measurements performed for various hydrocarbon substances irradiated at 77 K showed that absorption spectra cannot be approximated by a Gaussian distribution, which one would expect for electrons localized at identical capture centers. In the process of low-temperature radiolysis of organic compounds, the ejected electrons are stabilized at two different types of traps, *shallow* and *deep*, in competition with each other. Shallow traps are often likely to be permanent and intrinsic features of the polymer structure, probably "cavities" of various types formed by particular local configurations of the polymer chains, whereas deep traps are assumed to be radiation-produced free radicals. The density of electrons stabilized at shallow levels decreases with increases in the irradiation dose, reflecting preferential stabilization of electrons on free radicals at relatively high doses.

3.3.2 Luminescence Centers. Spectral Distribution of Thermoluminescence Emission

In many cases, a polymer does not itself give any significant luminescence and the observed emission is then from an impurity or additive molecules. Although the actual luminescence is due largely to chemical impurities, the charge trapping is due mainly to the basic polymer structure. The concentration of impurity molecules is often extremely low, and thus most of the initial ionization and excitation will occur in the polymer itself followed by charge and/or excitation energy transfer to the luminescence centers, causing these to excite. The question then arises as to how the energy available after charge recombination is transferred to these centers. One possibility is that both electrons and holes recombine within the polymer matrix and the energy is then transferred by an excitation mechanism to the luminescence center [20]:

$$M^+ + L + e^- \rightarrow M^* + L \rightarrow M + L^* \rightarrow M + L + h\nu \tag{3.4}$$

In Eq. (3.4), M and L are matrix and luminescence molecules, respectively, and the asterisk denotes an excited state. Alternatively, one of the charges migrates to and is trapped by the luminescence center, to be joined by the opposite charge when this is thermally detrapped:

$$M^+ + L + e^- \rightarrow M^+ + L^- \rightarrow M + L^* \rightarrow M + L + h\nu \tag{3.5}$$

A third possibility is that ionization occurring on the luminescence center results in the loss of a charge, which is trapped within the polymer matrix. This is then recaptured after thermal detrapping on warming:

$$L^+ + e^- + M \rightarrow L^+ + M^- \rightarrow M + L^* \rightarrow M + L + h\nu \tag{3.6}$$

Excitation transfer in polymers should be little affected by temperature, whereas charge-transfer distance may be small in many polymers at low temperatures (100 Å) but may increase very considerably at high temperatures [21].

The luminescence centers have been identified in a number of cases by comparison between the thermoluminescence emission spectra and known emission spectra. Different luminescent molecules (usually of aromatic nature) tend to be found in different structural regions and thus become associated with different glow peaks. Observation of various aromatic molecules is not too surprising, since contact with lubricating oils during manufacture is to be expected [22]. It is possible also that they result from atmospheric pollution. Air pollution studies have revealed many luminescent aromatic molecules [23], and these molecules could well get trapped within polymer samples after having diffused into them with the air [83,84]. Identification of luminescence centers can also be achieved by soaking the samples in a solvent and subsequently applying phosphorescence spectroscopy [22]. This process also may be used to introduce new, known luminescence centers into the polymer by keeping the samples for some time in a solution of the compound. Such a procedure, however, can induce essential structural changes in a polymer. Furthermore, the concentration of an impurity or a deliberate additive may change from bulk to surface, and this will occur in any species in which the binding energy to the surface differs from the binding energy to the bulk. Since dislocations present internal free surfaces, they too can be surrounded by impurity concentrations which differ from the bulk values [24,25].

Study of the spectral composition of the radiothermoluminescence provides information about both the polymer matrix itself and the impurities and additives dissolved in it which can form centers for the stabilization of charges and centers of luminescence. The bands in the radiothermoluminescence spectrum can be related to fluorescence or phosphorescence by comparison of the radiothermoluminescence spectra of the polymer with its fluorescence and phosphorescence spectra. For example, at 77 K, the luminescence of polyethylene consists mainly of phosphorescence with a maximum around 350 nm, whereas the luminescence spectrum of polypropylene consists of phosphorescence only [26]. The luminescence spectrum of polyethylene terephthalate at 170 K consists of phosphorescence (the long wave band) and fluorescence (the short wave band) [27]. The radiothermoluminescence spectrum of polystyrene consists of phosphorescence with a maximum around 420 nm and fluorescence in the region 300–350 nm [28].

The radiothermoluminescence spectra of polymers exhibit certain common features [29,80,81]. First, under isothermal conditions at 77 K, the luminescence spectra of polymers remain unchanged for at least an hour after the conclusion of irradiation. Furthermore, in the case of polyethylene, no change was observed in the luminescence spectrum for 30 min at 200 K after irradiation at 77 K and heating to 200 K [29].

Second, with increases in temperature, changes occur in the luminescence spectra. There is a change in the intensity ratio of the bands as well as a shift in the bands, which takes the form of a redistribution of luminescence increasingly toward the long wave bands.

Third, a large part of the radiothermoluminescence spectra of the polymers consists of long-wave luminescence (phosphorescence). This is particularly noticeable at low temperatures. Usually, there is a marked increase in the contribution of phosphorescence in radiothermoluminescence spectra as compared with photoluminescence spectra.

Fourth, increases in irradiation dose are accompanied by increases in the contribution of the long-wave part of the spectrum, and above a certain dose value, the spectral distribution of the radiothermoluminescence becomes independent of temperature.

The complex structure of the radiothermoluminescence spectrum and the changes in it with temperature can be explained in terms of a hypothesis postulating a large number of luminescence centers which are concentrated in different structural areas of the polymer and which become operative during heating of the specimen. The increase in the contribution of fluorescence in the spectra of some polymers (polyethylene, polyethylene terephthalate) on heating is most probably due to temperature quenching, which (as was shown in [30]) very markedly reduces the intensity of phosphorescence as opposed to fluorescence with increases in temperature.

Many changes in the radiothermoluminescence spectra can also be explained if the recombinational nature of the light emission in irradiated polymers is taken into account. This fact permits a simple explanation of the increase in the contribution of phosphorescence in recombinational luminescence as opposed to photo-stimulated luminescence. In the latter case, triplet states are formed mainly as a result of intersystem crossing, that is, $S_1 \rightsquigarrow T_1$, and therefore, the ratio of phosphorescence to fluorescence is always less than unity [31]. In recombinational luminescence, there are no limitations on the spin orientation of the recombining ions, and therefore, direct formation of excited molecules in the triplet state is possible. Furthermore, since the probability of favorable spin orientations of the recombining ions for the formation of a triplet state is three times greater than for the formation of a singlet state [32], phosphorescence can be expected to predominate over fluorescence in the case of recombinational luminescence.

Since the formation of excited molecules is due to energy released in the recombination of charges, the probability of the formation of an S_1 or T_1 state depends on the amount of energy. The amount of energy released is determined by many factors [33], including the depth of stabilization of an electron (the greater the depth of the trap for the electron, the smaller will be the energy of recombination). Thus changes in the depth of stabilization of electrons in irradiated polymers can cause changes in their radiothermoluminescence spectra. For example, redistribution of the luminescence increasingly toward the longer wave bands in the radiothermoluminescence spectra with increases in temperature is usually attributed to an increase in the percentage of recombining electrons which are stabilized at the deeper traps [34]. The intensification of fluorescence in the radiothermoluminescence of certain polymers also may be due to the fact that during heating some of the electrons are freed and recombine in the free state. This will cause an increase in the energy of recombination and a corresponding increase in the probability of the formation of molecules in the excited singlet state.

Changes in the spectral distribution of the radiothermoluminescence of polymers with increases in the irradiation dose are most probably connected with the accumulation of neutral products of radiolysis. The shift in distribution of stabilized charges toward deeper traps with increases in irradiation dose has been observed for a whole series of organic compounds [35] and is due to the accumulation of free radicals. When most of the charges are stabilized in deep traps, the value of the energy liberated during recombination is inadequate for excitation of the singlet energy level of the luminescence center, and therefore, the corresponding short wave band disappears from the radiothermoluminescence spectrum. With increases in irradiation dose, there is also a discrimination in favor of the lower levels among the triplet levels of the luminescence centers. Illumination of pre-irradiated samples with the light releases the electrons from the shallow traps and leads to the same qualitative changes in the spectral distribution of the radiothermoluminescence emission as does an increase in irradiation dose.

Thus the depth of stabilization of the electrons and the recombinational nature of the luminescence in irradiated polymers define the structure of the luminescence spectra and the changes in them with temperature and irradiation dose.

3.4   Kinetic Aspects of Charge Recombination in Organic Substances

The radiothermoluminescence glow curve is essentially a plot of the rate of ion recombination in a material as a function of temperature, although not all ions present will give luminescence on recombination because some types of excited states will return to the ground state predominantly by radiationless transitions. The thermoluminescence intensity I at any time t can be expressed as

$$I = -\alpha(dn/dt) \tag{3.7}$$

where dn/dt is the ion recombination rate, and α is the *luminescence constant,* i.e., the probability of photon emission from each recombination.

The kinetics of ion recombination will usually be either first- or second-order depending on whether the rate-determining step is charge untrapping or charge recombination after untrapping. For first-order ion recombination, the recombination rate is just

$$dn/dt = -Kn \tag{3.8}$$

where n is the number of trapped ion pairs at time t, and K is the recombination rate constant. The recombination rate constant is expected to depend, via a Boltzmann factor, on temperature:

$$K = S \exp(-E/kT) \tag{3.9}$$

where S is the frequency factor (sec^{-1}), and E is the "trap depth." Combining Eqs. (3.7)–(3.9), one gets

$$I = \alpha Sn \exp(-E/kT) \tag{3.10}$$

The factor α can be set equal to unity without any loss of generality as long as it remains constant with temperature. The solution of Eq. (3.10) gives, for thermoluminescence intensity, assuming $\alpha = 1$ and using linear heating scheme:

$$I = n_0 S \exp(-E/kT) \exp\left[-(S/\beta)\int_{T_0}^{T} \exp(-E/kT')\, dT'\right] \tag{3.11}$$

where n_0 is the initial concentration of the trapped electrons, T_0 is the initial temperature, and β is the constant heating rate, that is, $T = T_0 + \beta t$.

Equation (3.11) was deduced in the pioneering work of Randall and Wilkins [36]. The assumptions implied in Eq. (3.11) are as follows:

1. The concentration of free electrons is very small compared with the concentration of trapped electrons or ionized (empty) luminescence centers. Otherwise, it would be necessary to write

$$I = -\alpha(dm/dt) \tag{3.12}$$

   where m is the concentration of empty luminescence centers.

2. There is negligible retrapping of electrons; i.e., an electron released from a trap recombines with an empty luminescence center without first being retrapped. This is equivalent to stating that the luminescence process follows first-order kinetics with respect to the trapped electron concentration. However, if, for example, the probability of recombination equaled the probability of retrapping, the reaction would follow second-order kinetics.

3. All the traps have the same activation energy.

4. The thermally stimulated process leading to electron liberation follows the usual exponential inverse temperature behavior characteristic of an activated process.

In the case of second-order kinetics, the recombination rate and recombination rate constant are, respectively [68]:

$$dn/dt = -Kn^2 \tag{3.13}$$

and

$$K = S' \exp(-E/kT) \tag{3.14}$$

where S', the pre-exponential factor, is a constant with dimensions of cubic centimeters per second, which should therefore not be referred to as the "frequency factor."

From Eqs. (3.13) and (3.14), one gets

$$I = \alpha S' n^2 \exp(-E/kT) \tag{3.15}$$

Assuming that $\alpha = 1$, the solution of Eq. (3.15) is

$$I = n_0^2 S' \exp(-E/kT) \left[1 + (n_0 S'/\beta) \int_{T_0}^{T} \exp(-E/kT') \, dT' \right]^{-2} \tag{3.16}$$

The ion recombination rate constant increases with temperature (time), while the number of particles taking part in the reaction decreases. Thus the number of photons emitted per second reaches a maximum due to these two competing effects at a certain temperature T_p. At the point of maximum, the well-known condition $d^2n/dt^2 = 0$ should be fulfilled. Thus for first-order kinetics, the temperature of the glow peak maximum can be obtained by differentiation of Eq. (3.10), which for a linear warming rate of β gives

$$SkT_p^2/E\beta = \exp\ (E/kT_p) \tag{3.17}$$

An interesting feature, unique to the first-order case, results from Eq. (3.17). The initial concentration n_0 does not appear in this equation; therefore, the first-order peak is not expected to shift with various doses of irradiation.

For second-order kinetics, the maximum position may be expressed as follows [79]:

$$\beta/n_0 = \frac{2kT_p^2 S'}{E}\ \exp\ (-E/kT_p) - S' \int_{T_0}^{T_p} \exp\ (-E/kT)\ dT \tag{3.18}$$

Differently from first-order kinetics, the position of the maximum depends in this case on the initial charge concentration. The shift in T_p should be the same whether the heating rate is increased by a certain factor or the initial concentration of charges is decreased by the same factor.

In all cases studied so far, i.e., polyethylene [37,38], polybutadiene [38], polymethyl methacrylate [39], polystyrene [40], and squalene [20], the kinetics of the radiothermoluminescence appear to be first-order. Characteristically, the shape of the glow curve of a polymer is independent of the dose (up to 1 Mrad), although the radiothermoluminescence intensity increases with dose in the dose range 10^4–10^6 rads. The increase in the luminescence intensity with dose correlates with the known fact that charge concentration in low-temperature radiolysis increases linearly with the dose at doses lower than 1 Mrad. At a low dose, the distribution of stabilized charges in the bulk of the irradiated solids is not uniform. Most electrons are trapped within the Coulomb field of a parent ion after ionization, and on untrapping, they recombine with that ion. Thus the stabilized charges are distributed in pairs or groups, each recombination event occurring within an isolated microsystem, and the reaction follows first-order kinetics.

With increasing dose, the distance between individual pairs or groups of charges decreases and the distribution of ions becomes more uniform. When charges are uniformly distributed, the rise of their initial concentration n_0 results in displacement of the maximum to the low-temperature range, and this effect has been observed experimentally [38]. The overlap of tracks (or overlap of the Coulomb radii of neighboring positive ions) starts from 1–3 Mrads and results in the second-order recombination.

One of the ways of analyzing a thermoluminescence peak obtained using a linear heating function is by considering its symmetry properties. If it is remembered that a second-order glow curve arises because of the increased probability of retrapping, as opposed to the first-order case (i.e., the light emission is "delayed"), then it might be expected that a second-order peak will display more thermoluminescence during the second half of the peak than will a first-order peak [85]. The increasing part of the peak is governed in both cases by terms of the form $a \exp\ (-E/kT)$. However, for a first-order peak, the decreasing part follows

$\exp\left[-a'/\beta\int\exp\left(-E/kT\right)\,dT\right]$, as opposed to the second-order case, which follows $\left[a''+(a'''/\beta)\int\exp\left(-E/kT\right)\,dT\right]^{-2}$, where a, a', a'', and a''' are constants. This gives the first-order peak its characteristic asymmetry, where the half-width of the low-temperature side of the peak is almost 50% bigger than the half-width toward the fall-off of the glow peak [41]. Second-order peaks are characterized by a practically symmetrical peak.

A polymer usually has various kind of traps, and consequently, the glow curve exhibits several peaks. As long as these peaks are far apart, each of them can be considered to be a single peak. The situation becomes more complicated when two or more peaks overlap. The method for estimating the number and position of individual thermoluminescence peaks within a complex glow curve was preposed by McKeever [42]. A previously irradiated sample is first heated at a linear rate to a temperature T_{stop} corresponding to the position of the low-temperature tail of the first glow peak. The sample is then cooled rapidly down to the temperature at which irradiation was carried out and then is reheated at the same linear rate in order to record all of the remaining glow curve. The position of the first maximum in the glow curve is then noted. The whole process is repeated (on a freshly irradiated sample) using a different value of T_{stop} each time. If T_{stop} is increased in small increments and the corresponding T_p value is noted on each occasion, a plot of T_p against T_{stop} can then be made. For a glow curve that consists of one first-order peak only, the T_p–T_{stop} curve will simply be a line of slope 0 (Fig. 3.2a). This is so because the peak has not moved as the initial population of trapped charges is decreased. For a glow curve containing several well-separated first-order peaks, the T_p–T_{stop} curve

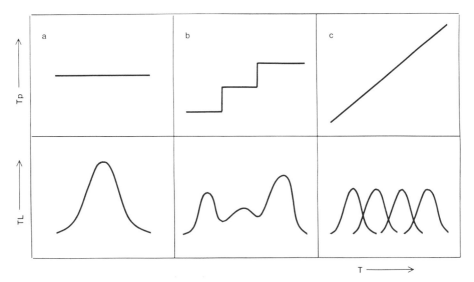

Fig. 3.2. Schematic T_p–T_{stop} curves for first-order thermoluminescence peaks. (*a*) A single thermoluminescence peak produces a straight line of slope 0. (*b*) A series of well-separated glow peaks produces a characteristic ''staircase'' structure. (*c*) Closely overlapping, or a quasi-continuous, distribution of thermoluminescence peaks produces a smooth line of slope ~1.0. After [42], by permission of Akademie-Verlag.

takes a "staircase" structure, with each flat region in the curve indicating the presence of an individual peak (Fig. 3.2b). For a series of more closely overlapping peaks, the T_p–T_{stop} curve becomes smoother, and the T_p values corresponding to the flat regions can only be used as indicators of the position of the individual peaks. As the peaks overlap more (Fig. 3.2c), the situation is tending toward the existence of a quasicontinuous distribution of peaks. In this case, the T_p–T_{stop} curve will have become a continuous line of slope ≈ 1.0. The limit of detection of flat regions within the T_p–T_{stop} curve gives a measure of the resolution of the technique, which is estimated to be approximately 5°C.

It has to be emphasized that the T_p–T_{stop} curves reveals only the most prominent of the peaks—smaller peaks remain hidden by their neighbors. Furthermore, only estimates of the true peak positions are given—the T_p values are usually slightly higher than the actual positions because of peak overlap.

Being based on a thermally agitated rate process, radiothermoluminescence may be expected to follow the time dependence of relaxation parameters and temperature for a simple relaxation process (single relaxation time):

$$\tau = \tau_0 \exp (E/kT) \tag{3.19}$$

Indeed, Eq. (3.19) is similar to Eq. (3.9). For transitions of a complex nature, however, the single-relaxation-time approach is generally found not to be valid. Williams, Landel, and Ferry have proposed an empirical equation which describes the temperature dependence of relaxation times in the glass-transition region [43]. This equation is now well known as the WLF equation and may be written generally as follows:

$$\tau = \tau_0 \exp \left[-\frac{C_1(T-T_g)}{C_2+T-T_g} \right] \tag{3.20}$$

where τ_0, C_1, and C_2 are constants, T is the temperature, and T_g is the glass-transition temperature of the material. By analogy, Eq. (3.9) can be replaced with an equivalent WLF form:

$$K = S \exp \left[\frac{C_1(T-T_g)}{C_2+T-T_g} \right] \tag{3.21}$$

This leads to the equation for thermoluminescence intensity:

$$I = \alpha Sn \exp \left[\frac{C_1(T-T_g)}{C_2+T-T_g} \right] \tag{3.22}$$

As before, the glow peak maximum T_p can be obtained by differentiation of Eq. (3.22), which for a linear warming rate of β gives

$$\frac{\beta C_1 C_2}{S[C_2+(T_p-T_g)]^2} = \exp \left[\frac{C_1(T_p-T_g)}{C_2+(T_p-T_g)} \right] \tag{3.23}$$

This equation can be used for evaluation of the frequency factor S, although the value $(T_p - T_g)$ may be subject to considerable error because of uncertainty in the value of T_g [44].

3.5 Isothermal Luminescence: Tunneling and Diffusion-Type Charge Recombination

When the luminescence emission of a polymer is measured during and following irradiation, it can be seen that a sudden rise in light intensity occurs as soon as high-energy radiation impinges on the sample. Thereafter, it slowly approaches a plateau indicative of a steady state. When irradiation is turned off, the emission at first drops sharply and then decays steadily for up to several days [45].

The initial rapid increase in luminescence is due to fluorescence and phosphorescence, as well as to Cerenkov radiation. The latter electromagnetic radiation is produced whenever a charged particle passes through a medium in which the phase velocity of light is less than the particle velocity, that is, when $\beta n > 1$ (n denotes refractive index, $\beta = v/c_0$). The input of Cerenkov's radiation increases both with increases in the energy of irradiation and with increases in the refractive index of the material being irradiated. Cerenkov photons are expected to make a significant contribution to the emission of light during irradiation because of the estimated low photon yield (about 10^{-3}) for ion recombination.

The luminescence given off during irradiation at 77 K will be caused by prompt recombination as well as, in part, by the isothermal decay. The luminescence intensity under these conditions must be related to the rate of production of electrons, which is proportional to dose rate and the rate of return of electrons to positive ions. As soon as irradiation of the sample commences, thermal contact of the electrons with the frozen matrix will be initiated. At first, the majority of traps, those believed to be characteristic of the rigid polymer, will be filled, and later, as more electrons are produced, thermal and photoequilibria will be established with all the traps of the polymer. At this point, constant luminescence intensity is reached which is proportional to the steady-state concentration of the mobile electrons. Provided that the yield of radiation-induced products with a positive electron affinity is relatively small, the concentration of all transient intermediates directly or indirectly involved in the luminescence will remain time-independent.

The isothermal luminescence decay following exposure to low doses (10^4–10^5 rads) is not measurably dose-dependent [45]. This suggests that each recombination occurs as an isolated event between geminate, or correlated, charge pairs. First-order kinetics then apply. From Eq. (3.8),

$$n = n_0 \exp(-Kt) \tag{3.24}$$

and since the luminescence intensity $I \propto dn/dt$,

$$I = I_0 \exp(-Kt) \tag{3.25}$$

At higher doses, the half-life of the decay becomes dose-dependent. This can be interpreted by assuming a considerable degree of charge randomization, which leads to second-order recombination. From Eq. (3.13),

$$n = n_0/(1 + n_0Kt) \qquad\qquad (3.26)$$

so n decreases hyperbolically with time. The luminescence decay in this case is given by

$$I = I_0/(1 + \sqrt{I_0K}t)^2 \qquad\qquad (3.27)$$

Simple exponential decay and a power-law decay $I \propto t^{-2}$ which one would expect, respectively, for first- and second-order recombination kinetics with a single trap depth are observed, however, quite seldom. More frequently, the isothermal luminescence follows an unusual time relation, in that light intensity decreases with time as $1/t$ [5,46,82].

An illumination of the irradiated sample with light during the isothermal decay causes an essential increase in the emission of light during illumination and shortly thereafter. The decay eventually, however, resumes the pace it would have followed had no bleaching taken place. This same effect can be achieved by the application of an electrical field [47]. While every consecutive illumination with light procures a significant increase in light emission, the first application of the electrical field produces a considerably greater effect than consecutive applications. However, if the field is reversed and reapplied, luminescence enhancement can again be very large. This observation clearly shows that as opposed to enhancement with light, application of an electrical field involves a directional effect in the enhanced rate of electron untrapping which ultimately accounts for the luminescence. The model to consider is a series of electrons distributed at various distances from their respective cations and in random directions relative to the external field. Thus with one field direction electrons lying on one side of their cations will be encouraged to recombine with them, whereas the electrons lying on the other side will find their untrapping inhibited until the field is reversed.

It is reasonable to assume that both illumination with light and application of an electrical field cause the mobilization of electrons from deeper traps which were less, or not at all, accessible to recombination during the isothermal decay. It also appears that no significant change in the population of electrons trapped in the wells which are emptied during the decay occurs.

Let us now consider two methods of electron release: thermal untrapping and tunneling. The problem in separating these two possibilities is that interpretation of thermoluminescence decay is so complex that the decay can be used to infer the specific mechanism or processes involved only with essential reservations, since each individual result can (by suitable assumptions about distributions and filling of traps) be interpreted in terms of different mechanisms. For instance, the intensity of isothermal recombinative luminescence ($I \approx 1/t$) can be derived on the basis of both electron tunneling and diffusion-type charge recombination.

If there is a uniform distribution of trap levels, i.e., equal number of traps at all depths, then the luminescence decay is given by [48]:

$$I = \int_0^\infty N_E \, S \exp\left(-E/kT\right) \exp\left[-St \exp\left(-E/kT\right)\right] dE \tag{3.28}$$

where N_E is the number of traps between E and $E + dE$. Upon integration, Eq. (3.28) yields:

$$I = n_0 kT/t \tag{3.29}$$

assuming that $st \gg 1$. Thus a uniform distribution of trap energies will yield the t^{-1} law directly.

An alternative theory is that electrons trapped in the vicinity of a cation can tunnel to it with little or no temperature dependence. The rate constant of this process varies exponentially with the distance separating the reacting pair [49]. The overall process is determined by the disappearance of pairs differing by their distance of separation. The formal kinetic equation for the time dependence of the concentration of reacting charges can be written as [86]:

$$n/n_0 = 1 - k_2 \ln (t/t_0) \tag{3.30}$$

where k_2 is the effective rate constant. Differentiating Eq. (3.30) and considering that the intensity of luminescence is proportional to the rate of the process dn/dt [Eq. (3.7)], an expression discribing the evolution of luminescence intensity with time is

$$I = n_0 k_2/t \tag{3.31}$$

Thus allowing for the coulombic potential of the cation, the inverse time dependence can be explained assuming a single trap depth with electrons at various distances from their respective cations.

An attempt to distinguish between traps involved in tunneling and thermal untrapping in irradiated polyethylene has been undertaken recently by Charlesby [5]. He concluded that the former traps are deeper but are located closer (60 Å) to the parent ions and that electrons which fill them are primarily geminate electrons. The remaining traps are located at random within the specimen and are more shallow. These traps can be filled at random by more mobile (i.e., nongeminate) electrons distributed throughout the specimen. Recombination following release can then occur with any ion encountered during subsequent wandering of these electrons within the specimen. Some of the electrons released on warming may be recaptured in one of the deeper traps near a cation and thus provide an additional contribution to tunneling when the temperature is reduced.

The preceding explanation, however, contradicts the well-known experimental facts. The assignment of the thermal untrapping to the recombination of randomly distributed nongeminate electrons requires second-order reaction kinetics at relatively high temperatures, where recombination is highly temperature-dependent and the tunneling contribution cannot be large. Also, since the geminate/nongeminate electron ratio is about 40 [71], higher thermoluminescence output at low temperatures is to be expected, and this is not the case in many instances.

It is more probable that both tunneling and thermal untrapping contribute to charge recombination in polymers, depending on the temperature. In the vicinity of the temperature of a relaxation transition, when the intensity of molecular motion in the polymer is high, the distribution of charges will be governed by their diffusion and the tunneling contribution can be disregarded [50]:

$$I_d \gg I_t \quad \text{and} \quad I = I_d \tag{3.32}$$

where I_t and I_d are the isothermal recombinative luminescence intensities due to tunneling and diffusion, respectively. At sufficiently low temperatures, when molecular motion in the polymer is frozen, the diffusion mechanism of charge destruction can be disregarded and charge recombination will be controlled by tunneling:

$$I_t \gg I_d \quad \text{and} \quad I = I_t \tag{3.33}$$

In general, one can write the following expression for the rate of charge recombination:

$$I = I_t + I_d \tag{3.34}$$

While the tunneling contribution is temperature-independent, the diffusion contribution is highly temperature-dependent:

$$I_d = I_0 \exp\left(-E/kT\right) \tag{3.35}$$

where E is the activation energy for charge diffusion, i.e., the activation energy for molecular mobility in a given relaxation transition. Conditions (3.32) and (3.33) can be regarded as defining the high and low temperatures for a given relaxation transition. The low temperature for one transition may be the high for another, depending on the activation-energy value.

3.6 Variation of Luminescence Intensity with Dose and Dose Rate

The thermoluminescence light output essentially is a measure of the number of trapped ion pairs in the material that are capable of producing emission.

The most common dose curves increase linearly to a maximum, often at 1–3 Mrads, and then decrease steadily as the dose increases [14,38]. The linear increase in glow intensity with increasing dose observed at low doses indicates that the electron traps are already present within the polymer and not generated by the radiation.

If they were, the glow intensity would be expected to increase superlinearly with dose, since the increasing concentration of traps would give a higher probability of free-electron trapping. The actual decrease in thermoluminescence intensity with dose could be due to destruction of traps, destruction of luminescente molecules, or production by radiation of species which compete for charges or accept excitation energy from excited luminescent molecules [11]. Among these possibilities, production of charge traps by radiation, linked with radiation untrapping, seems to be the most probable reason for the intensity decrease at high doses and is related to the production of free radicals by the radiation.

Partridge proposed a model in which it is assumed that the ejected electrons are captured by two different types of traps, *shallow* and *deep,* in competition with each other [51]. Shallow traps are "cavities" of various types, whereas deep traps are presumably radiation-produced free radicals. Shallow traps, on warming, release their electrons back to the ionized luminescent molecules, whereas deep traps remove the electrons completely from the thermoluminescence process. If radiation untrapping affects only electrons in the shallow polymer traps, then the thermoluminescence intensity will decrease at high doses, since the radicals capture most of the new electrons while the electrons already in shallow traps are steadily untrapped by the radiation.

The mathematical treatment of this model has been given for a system subjected to UV [52] and true ionizing [11,51] radiation. In the latter case, the following set of equations has been introduced;

$$i + f = Hr \tag{3.36}$$
$$di/dr = GEf/(D + Ef) \tag{3.37}$$
$$ds/dr = GD/(D + Ef) - Fs \tag{3.38}$$

where s, f, and i are the concentrations of shallow-trapped electrons, free radicals, and radical-ions, respectively, at dose r; G and H are dose-independent rates of electron trapping (by all traps) and radical production, respectively; F is the rate constant for untrapping of shallow-trapped electrons by radiation; D is the probability of an electron being captured by a shallow trap; and Ef is the probability of its capture by a free radical. It is assumed that empty shallow traps greatly outnumber filled ones and that the rate of radical production changes linearly with dose.

The preceding set of equations gives the approximate relation for the dose r_{max} at which the thermoluminescence intensity passes a maximum:

$$(r_{max})^4 \simeq HD/0.3(H - G)EF^3 \tag{3.39}$$

As it follows from Eq. (3.39), the position of the maximum on the thermoluminescence intensity versus irradiation dose curve is determined by F and the D/E ratio, i.e., by the rate of radiation-induced untrapping and by the relative electron capture efficiencies of shallow traps and free radicals. For the same value of r_{max}, the thermoluminescence intensity will decrease more steeply at high doses if D/E and F are large, and vice versa.

Along with the thermoluminescence intensity versus irradiation dose curves which exhibit maxima, the other kinds of dependencies have been noticed. Boustead and Charlesby observed saturation of the thermoluminescence intensity for some peaks at high doses [20,53]. If I is the intensity of the glow peak maximum, r is the dose, and I_0 and A are constants, the change in intensity of the thermoluminescence peak with dose can be presented as

$$I = I_0[1 - \exp(-Ar)] \tag{3.40}$$

The most probable reason for this type of curve is untrapping of trapped charges by the radiation, since polymer thermoluminescence is activated by molecular chain motion and a considerable amount of the irradiation energy is likely to be ultimately dissipated in molecular vibrations.

A third, less common type of thermoluminescence intensity versus dose curve can be roughly represented as

$$I = Br(r + C) \tag{3.41}$$

where B and C are constants. This behavior, found in one glow peak of squalene and one of polyethylene [20,54], indicates creation of extra charge traps by radiation in addition to the traps already present.

Two last types of thermoluminescence intensity versus dose curves can be deduced as particular cases from more general behavior described by Eqs. (3.36)–(3.38). The exponential dependence of Eq. (3.40) and "linear plus quadratic" dependence of eq. (3.41) follow from Eqs. (3.36)–(3.38) if $E = 0$ and $F = 0$, respectively.

3.7 Factors Influencing Radiothermoluminescence

3.7.1 Presence of Oxygen

Oxygen dissolved in a polymer may have a significant effect on the glow curve, and its presence can be accompanied by the removal of some thermoluminescence peaks and the appearence of new ones. This effect has been most completely studied for polyethylenes. It has been established that dissolved oxygen has virtually no effect on the lowest temperature peak at 130 K but either completely removes peaks at 180 and 240 K (doses below 0.2 Mrad) or essentially reduces their intensity [55,78]. At the same time, a new peak is produced around 145 K, its temperature position being intermediate between that of the peaks at 130 and 180 K of the oxygen-free glow curve (Fig. 3.3). The oxygen effect can be removed by pumping or by a pre-irradiation followed by preheating. The latter method removes oxygen by conversion into peroxide radicals, which are produced when oxygen combines with alkyl radicals formed during the pre-irradiation. The peak around 145 K was found to occur only in the presence of oxygen molecules, but not with oxygen in the form of peroxides. The removal of the oxygen effect is permanent only as long as the polyethylene is kept from contact with oxygen. Specimens allowed to stand in air at room tempera-

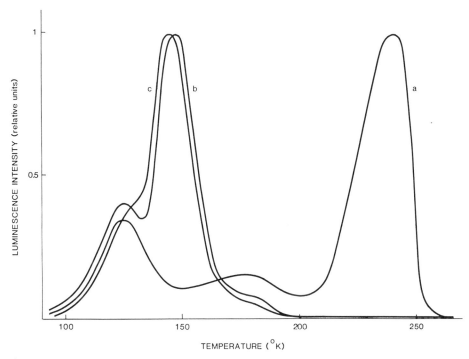

Fig. 3.3. Glow curves of polyethylene samples saturated with oxygen at pressures of (*a*) 0, (*b*) 150, and (*c*) 500 torr. After [78], by permission of Pergamon Journals, Inc.

ture reabsorb oxygen, and the glow curve of such specimens will be again characterized by the presence of the maximum around 145 K.

Two basic mechanisms describing the influence of oxygen on the character of radiothermoluminescence have been proposed. According to one of them (Nikolskii *et al.* [56]), oxygen becomes mobile as a result of the onset of a particular type of molecular motion, combines with the electron traps, and in so doing releases any trapped electrons. Electron traps are assumed to be alkyl radicals. Thus, in the presence of oxygen, the recombination of some charges occurs at temperatures lower than it would have in the oxygen-free sample. This mechanism is based on two experimental facts: first, the onset of "oxygen peak" is observed to take place at about the same temperature as oxygen molecules start to become mobile, and second, the main formation of peroxide radicals during warming occurs very suddenly at the exact temperature region of the "oxygen peak." A different possibility which also emphasizes oxygen diffusion but does not involve the formation of peroxide radicals has been described by Brocklehurst *et al.* [58]. According to these authors, oxygen becomes mobile before the large anions R^- (but after the electrons in physical traps); it diffuses through the matrix, absorbing the electrons from R^- to form O_2^- ions, which then neutralize the cations, giving modified yields of excited states.

The other mechanism (Charlesby, Partridge [13,37]) suggests that oxygen acts as an effective electron trap. Here, it is also assumed that the appearence of a new maximum results in a decrease in the thermoluminescence intensity at higher temperatures, but the redistribution of charges is believed to take place during the irradiation as a result of the formation of O_2^- ions. However, it is not clear in this hypothesis why the O_2^- molecules should not be formed at any temperature down to that of liquid nitrogen, since although the oxygen molecules may not be mobile in the polymer at these low temperatures, the electrons certainly are mobile. It is worthwhile also to mention the study carried out by Meggitt and Charlesby [57] on polyethylene samples soaked in a solution of biphenyl in hexane with subsequent hexane evaporation. The biphenyl anion concentration following irradiation was not found to be significantly reduced by the presence of oxygen in the sample, although this would be expected to compete with biphenyl for mobile electrons. However, at the 150 K peak associated with the presence of oxygen, the biphenyl anion concentration was found to sharply decrease. This suggests that oxygen, on becoming mobile, collects electrons from the anions.

The additional results supporting the mechanism of oxygen influence on the glow curve as being connected to the oxygen mobility were obtained by studying oxygen diffusion in polyethylene at 195 K [59]. Figure 3.4 presents the glow curves (200–270 K temperature interval) obtained for a series of polyethylene samples. Identical samples were irradiated in a vacuum at 77 K. Then some samples were heated to 195 K, kept at this temperature for some time, and cooled back to 77 K. It was found that the intensity of the maximum at 240 K depends on whether all operations following irradiation at 77 K were performed in vacuum or in air. Vacuum treatment practically did not change the intensity of the peak at 240 K. However, an essential decrease in the intensity of this peak was noticed for the samples exposed to the treatment in air. Oxygen influences an irreversible decrease (and eventually, disappearence) of the maximum at 240 K, since evacuation at 195 K of the samples treated in air and subsequent cooling to 77 K under vacuum were not accompanied by restoration of the original intensity of the maximum. Thus oxygen, diffusing in irradiated samples at 195 K, leads to the distruction of charges stabilized in the process of low-temperature radiolysis.

The intensity and position of the ''oxygen peak'' were found to depend on both oxygen concentration and irradiation dose [59]. Up to a certain oxygen concentration, the temperature of the maximum is constant and depends only on the irradiation dose. Further increases in oxygen concentration are accompanied by a shift in the position of the maximum to lower temperatures. This effect was noticed at different irradiation doses, but it was most profound at low doses.

Assuming that oxygen diffusion is the limiting stage of the reaction, two cases of relative oxygen and charge concentrations may be considered:

$$[O_2] \gg [R^-]$$
$$[O_2] \ll [R^-]$$

During the time of diffusion t, the temperature of the sample increases by $\Delta T = \beta t$, where β is the heating rate. When $[O_2] \gg [R^-]$, the lower the concentration of oxygen, the greater are the t and ΔT values. This explains the shift of the maximum with oxygen concentration; namely, at high oxygen concentrations, the temperature of the

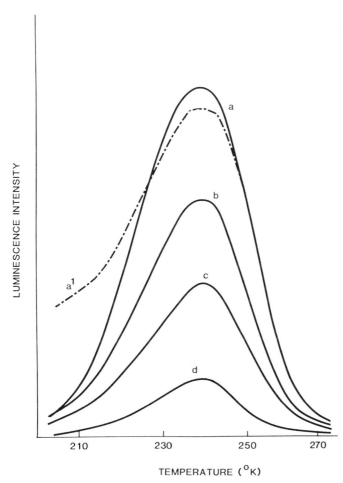

Fig. 3.4. Influence of oxygen present in air on the radiothermoluminescence of polyethylene: a^1 = original sample; a = sample kept in vacuum at 195 K for 120 min; b, c, and d = samples exposed to air at 195 K for 30, 60, and 120 min, respectively [59].

"oxygen maximum" decreases with increases in oxygen concentration. In the case of low oxygen concentration, the average distance between charges and oxygen molecules, and therefore, the diffusion time, does not depend on oxygen concentration. Thus, at low oxygen concentrations, the position of the maximum is independent of oxygen concentration and is defined by the irradiation dose.

3.7.2 Optical Bleaching

Organic substances irradiated at 77 K luminesce at this temperature under the action of visible light. When the light source is extinguished, an afterglow is observed, the duration of which depends on the material, the dose received during radiolysis, the duration of the exposure to light, etc. This phenomenon—*radiophotoluminescence*—has been observed in a variety of organic substances [4,14].

The result of illuminating the irradiated sample is to produce a substantial reduction in the coloration of the sample and in the light output (optical bleaching effect).

Ion recombination occurs under the action of light as long as the energy of the light quanta is large enough to free the captured electron from the trap. The electron will then be captured again in a trap or will recombine with a positive ion. When polymer samples are exposed to monochromatic light, the thermoluminescence intensity falls to a certain limiting value. The latter depends only on the wavelength of the incident light. Increases in intensity of the monochromatic light lead to a more rapid fall in the thermoluminescence intensity to the limiting value, but the limiting level remains unchanged.

Optical bleaching reduces the thermoluminescence intensity by different amounts in different materials, but in all cases a sudden dropoff in thermoluminescence is observed at the very beginning of bleaching. Bleaching is most easily accomplished when transparent samples of vitrifying materials, large crystals, and thin polymer films are used. It goes more slowly in opaque polycrystalline samples.

Along with decreasing the intensity of emitted light on subsequent heating, bleaching may produce significant changes in the electron-spin resonance spectra [3,14]. If the ions stabilized in the traps during radiolysis have unpaired spin, they will be recorded in the electron-spin resonance spectra along with the stabilized radicals. Paramagnetic centers of two types are formed during radiolysis, namely, radicals and centers, the paramagnetic absorption of which shows up in the electron-spin resonance spectrum as a single narrow line. As a rule, neither the form nor the intensity of the electron-spin resonance spectra of the irradiated material changes during optical bleaching as long as the irradiation dose is greater than 15–20 Mrads. However, in samples irradiated with smaller doses, bleaching produces appreciable changes in the spectra. The nature of the changes indicates that the electron-spin resonance spectrum of an irradiated sample must be regarded as the result of superimposing the spectrum of a radical on a single line in the center. During optical bleaching, the line in the center disappears. Buildup of the paramagnetic centers associated with the occurrence of such a single line in the electron-spin resonance spectrum is greatly slowed up at doses larger than several megarads. At any rate, the line in the center only distorts the radical spectrum in an unbleached sample at comparatively small doses. For example, the distortion of the spectrum of the alkyl radical in polyethylene only occurs at doses less than 1 Mrad. No changes were observed in the edges of the electron-spin resonance spectra during bleaching. This allows one to conclude that the concentration of radicals apparently does not change substantially during bleaching [14].

Low-temperature radiolysis leads to the stabilization of charges in the traps which have wide depth distribution. Optical liberation of the charges from these traps requires considerable energy (1–3 eV), since the concentration of charges captured in shallow traps (depth less than 1 eV) is very low [35].

In the majority of cases, samples that have been given an optical bleaching do not show any appreciable changes in the form of the thermoluminescence curve as compared with unbleached samples [4], although a preferential bleaching of the low- [60,61] and high-temperature [62] portions of the glow curve has been reported. A characteristic feature is that re-irradiation of a bleached, but still frozen sample completely restores the thermoluminescence and photoluminescence intensity in the sample. The experimental fact that bleaching does not change the shape of the luminescence curve is of a considerable interest. It indicates that throughout a wide temperature region (from liquid nitrogen temperature up to the glass-transition temperature), heating of a pre-irradiated polymer leads to recombination of the charges in the traps with the same depth distribution. This conclusion is contrary to the widely held opinion that the first stage of heating of the irradiated substance is accompanied by recombination of the charges captured only in shallow traps [63,64], where it is postulated that thermal liberation of the charges first from the shallow and then from the deeper traps precedes their recombination. However, since the irradiated polymers do not contain charges captured in the traps from which they could be released by thermal activation, already the initial low-temperature recombination involves charges stabilized in deep traps (2–3 eV).

Some of the experimental results seem to be in contradiction with the preceding conclusion. For example, Charlesby and Partridge [55] found that if, during warming, a sample is suddenly quenched back to the irradiation temperature, then, on subsequent warming, a little light is emitted until just below the quenching temperature, and this indicates that shallow traps are emptied before deep ones. A likely explanation of this situation is that the traps, of all types, are distributed at different distances from the polymer chains [11]. Thus at low temperatures only charges trapped close to a chain (or near a particularly mobile segment of a chain) are released, whereas at higher temperatures charges farther away can get untrapped as the frequency and amplitude of the molecular motion increase. Thus the different glow peaks are associated more with different frequency constants than with different activation energies [37]. Aulov et al. [62] carried this temperature quenching a stage further by irradiating polyethylene at 4 K, warming it to 180 K, quenching it back to 4 K, and then bleaching with visible light. On subsequent warming, some luminescence did appear in the 4–180 K range, which indicates that while the untrapping caused by optical bleaching usually results in ion recombination, some of the released charges are trapped again in other traps.

Bleaching not only takes place as a result of illumination of the irradiated material with visible light, but it may also accompany the radiothermoluminescence experiment even when it is conducted in complete darkness. The intensity of light measured at a particular temperature is proportional to the rate of electron recombination [see Eq. (3.7)]. A fraction α of the electrons causes an emission of light. Most electrons contributing to the luminescence of the irradiated polymer are mobilized through an erosion of traps. A fraction, however, is generated by light-induced detrapping. Photons produced by the former process cause the liberation from traps of more electrons, which, in turn, can cause luminescence. The secondary photons will bleach more electrons, and so forth.

The total luminescence intensity is then [19]:

$$I(T) = -(dn/dt)_e(\alpha + K\alpha^2 + K^2\alpha^3 + \cdots) \tag{3.42}$$

where $(dn/dt)_e$ is the rate of electron mobilization by erosion of traps, and K is the probability that a photon will remove an electron from its trap. The bleaching at a temperature T will free electrons from traps deeper or of greater electron affinity than those which are eroded at T. Hence the more bleaching that occurs at or before T, the smaller are the luminescence intensities at temperatures above T. However, the overall contribution of bleached electrons to the luminescence given off by the polymer should be relatively small. This follows from the fact that α and K are both smaller than 1.

3.7.3 Temperature Quenching

As a general rule, luminescence increases in efficiency as the motion of a molecule is restricted, since the competing processes of radiationless energy transfer require coupling between the excited molecule and the molecules which surround it, and this coupling becomes greater with increased amplitude and diversity of molecular motions. Lowering the temperature, therefore, usually increases the probability of luminescent processes. Quenching of phosphorescence in low-molecular-weight organic scintillators has been known for some time [65], and a similar effect should be expected in polymers. An increase in intersystem crossing probability, and thus a decrease in fluorescence emission, may result from the perturbation due to increasing temperature [66]. In a similar way, the perturbation due to an impurity molecule could increase the intersystem crossing probability or could result in the recombination process leaving the excited recombination species in a triplet state; both processes would cause decreased fluorescence and increased phosphorescence. If a major part of the thermoluminescence is basically phosphorescence emission from excited molecules formed by ion recombination, such phosphorescence can be quenched by thermally activated collisions between the excited molecules and the polymer chains. Indeed, it is expected that in the case of thermoluminescence induced by radiation, the triplet/singlet ratio is of about 3:1 [67], and the same molecular motion may be simultaneously promoting thermoluminescence by causing ion recombination and reducing it by phosphorescence quenching.

Obviously, the nature of the molecules responsible for the light emission is of importance. If the luminescence center is solely a fluorescence emitter, such as anthracene, for instance, then the intensity of the emission is virtually independent of temperature, and the principal factor governing the intensity of thermoluminescence is the rate of liberation of electrons, which decides the form of the glow peak. However, when the luminescence centers are solely phosphorescence emitters, such as benzophenone, the fraction of excited molecules which decay radiatively will be affected by temperature, since phosphorescence emission is extremely sensitive to collisional deactivation—a temperature-dependent phenomenon. The glow curves of n-heptane doped with anthracene and benzophenone are shown in Fig. 3.5. Curve 3 of this figure shows the measured phosphorescence intensity (in relative units) of n-heptane doped with benzophenone under UV excitation with the phosphorescence intensity at 77 K set equal to unity [53]. As can be seen, curve 2 is the product of curves 1 and 3. Thus a glow peak produced by n-heptane doped with benzophenone is a product of two factors—the rate of untrapping of electrons and the rate of collisional quenching after an electron has combined with a luminescent ion. Boustead and Charlesby [69] showed that a similar quenching mechanism occurs in the ther-

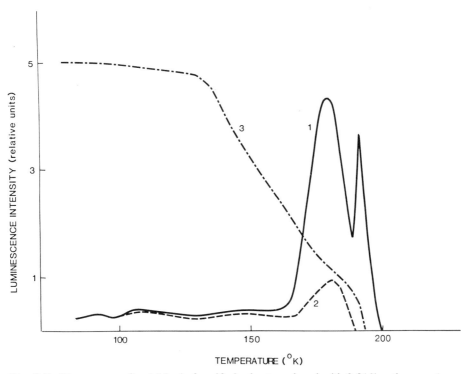

Fig. 3.5. Glow curves after 1 Mrad of purified *n*-heptane doped with 0.01% anthracene (curve 1) and benzophenone (curve 2). Curve 3 shows the UV-stimulated phosphorescence intensity of the sample doped with benzophenone as a function of temperature. After [53], by permission of Royal Society, London.

moluminescence of polyethylene and low-molecular-mass alkanes. These authors considered a system in which a small number of excited phosphorescent molecules are embedded in a nonphosphorescent matrix. Depopulation of the lowest triplet state of the luminescent molecules may be either radiative, giving rise to phosphorescence, or nonradiative, in which case the excess energy is dissipated thermally. If rate constants for these two processes are k_p and k_n, respectively, then the quantum efficiency for phosphorescence q_p is given by

$$q_p = k_p/(k_p + k_n) \tag{3.43}$$

The nonradiative constant, k_n, can be written in terms of a temperature-dependent component $(k_n)_T$ and temperature independent component $(k_n)_0$:

$$k_n = (k_n)_0 + (k_n)_T \tag{3.44}$$

The constant $(k_n)_T$ is due primarily to collisional deactivation and could be written in the form of a Boltzmann function:

$$(k_n)_T = \epsilon \exp\ (-W/kT) \tag{3.45}$$

where ϵ is a constant, and W is the activation energy of the quenching process.

Combining Eqs. (3.43)–(3.45), the phosphorescence quantum efficiency becomes

$$q_p = k_p/[(k_n)_0 + k_p + \epsilon \exp\ (-W/kT)] \tag{3.46}$$

Replacing quantum efficiency with the luminescence constant α_T at temperature T, an expression for α_T can be derived from Eq. (3.46):

$$1/\alpha_T - 1/\alpha_0 = a \exp\ (-W/kT) \tag{3.47}$$

where the various constants of Eq. (3.46) are absorbed into the constant a, and α_0 is the luminescence constant at zero collisional quenching. It is assumed to be sufficiently accurate to replace α_0 by α_{77}, the luminescence constant at liquid nitrogen temperature, to give

$$1/\alpha_T - 1/\alpha_{77} = a \exp\ (-W/kT) \tag{3.48}$$

Since the luminescence intensity is proportional to luminescence constant, Eq. (3.48) can be expressed by utilizing some constant a' instead of a:

$$1/I_T - 1/I_{77} = a' \exp\ (-W/kT) \tag{3.49}$$

Application of Eq. (3.49) to alkanes and polyethylene samples gave good agreement with experiments, and it was concluded that quenching is independent of the nature of the impurities and is a function solely of the polymer matrix [69].

3.7.4 Electric Field Effects

It was reported by several researchers that isothermal thermoluminescence at low temperatures prior to warming is affected by an external electric field. Similar effects were observed for inorganic [70] and organic [71] solids.

The effect of an electric field of 2.5×10^5 V cm^{-1} on the isothermal thermoluminescence of high-density polyethylene at 87 K exposed to 1 Mrad of γ-radiation was studied by Blake and Randle [72]. The increases in light intensity at A and B (Fig. 3.6) correspond to short applications of this field. They show the same fractional increase relative to the isothermal thermoluminescence (the same enhancement ratio). The field was reapplied at C, which produced the same enhancement ratio as at A and B, but the light intensity decreased during prolonged application of the field. Between C and D, the field was removed and reapplied three times with a consequent reduction in the enhancement ratio. The field polarity was reversed at E when the initial enhancement ratio was found to be equal to that of A, B, and C. Thus the reversal of the field polarity stimulated additional luminescence from the sample.

It would seem that postirradiation electroluminescence arises from the same population of trapped positive ions and electrons as does isothermal thermoluminescence,

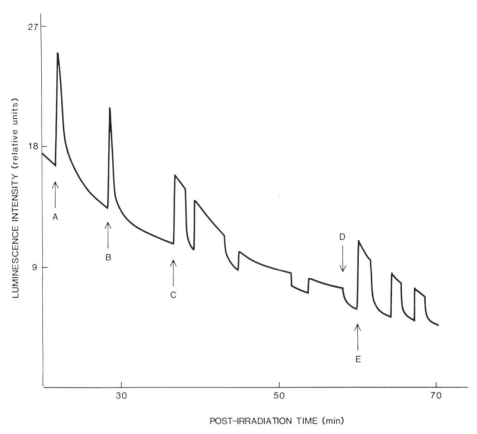

Fig. 3.6. The effect of an applied electric field on the isothermal thermoluminescence decay at 87 K in high-density polyethylene. After [72], by permission of the Institute of Physics, Bristol, England.

since there is a fixed ratio between the electroluminescence signal and isothermal thermoluminescence during the decay of the latter. It was concluded that field-assisted thermal detrapping is responsible for the effect.

The observation that both polarities produce more or less the same enhancement of thermoluminescence intensity and their effects are largely independent suggests that one can crudely divide the trapped electron population into two groups. However, the action of the applied field alone cannot explain the two-component nature of the results. If electrons are trapped outside the Coulomb well of their positive ions, then the applied field would always lead to an enhanced emission, and field reversal would not produce a further increase, as shown at E in Fig. 3.6. However, if the electrons are trapped closer to the positive ions, where the Coulomb field is comparable with the applied field, two-component behavior would arise as the field either assisted or hindered the positive ion field. On average, half the trapped electrons will lie on one side of their positive ions and half will lie on the opposite side as related to the direction of the applied electric field ϵ. One group will be more rapidly detrapped, since the activation energy is decreased by an amount $ae\epsilon$, where a is the trap half-width, and e is the electron charge. The latter group will find detrapping more difficult, because their activation energy is increased by a similar amount. Assuming that isothermal thermoluminescence decay is a simple exponential, the trapped electron lifetime τ can be related to the trap depth E by

$$\tau = \tau_0 \exp (E/kT) \qquad\qquad\qquad\qquad\qquad (3.50)$$

where τ_0 is related to the period of lattice vibration. The trapped electron population decays according to

$$-dn/dt = n/\tau \qquad\qquad\qquad\qquad\qquad (3.51)$$

and the light emission decays according to

$$I = -\alpha(dn/dt) \qquad\qquad\qquad\qquad\qquad (3.52)$$

where α is a constant.

The effect of the external field ϵ is to split the electron population by raising the activation energy of one half to $(E + ae\epsilon)$ and lowering the other half to $(E - ae\epsilon)$. If the subscripts s and f are used to distinquish the two groups, we have

$$\tau_s = F\tau \qquad \text{and} \qquad \tau_f = \tau/F \qquad\qquad\qquad\qquad\qquad (3.53)$$

where $F = \exp (ae\epsilon/kT)$.

The equations for two populations now become

$$n_s = \tfrac{1}{2}n_0 \exp (-t/F\tau) \qquad \text{and} \qquad n_f = \tfrac{1}{2}n_0 \exp (-tF/\tau) \qquad\qquad (3.54)$$

where n_0 is the total electron population at $t = 0$.

The total emission in the presence of the field is then given by

$$I(\epsilon, t) = (\alpha/\tau)[(n_s/F) + Fn_f] \qquad\qquad\qquad\qquad (3.55)$$

Thus the original single-exponential isothermal thermoluminescence decay has been split into two exponential components whose magnitudes and decay rates are dependent on the external field and the width of the electron trap. At $t = 0$, the ratio of intensities with and without the external field becomes simply $\cosh (ae\epsilon/kT)$.

The model described above accounts for many features of the observed experimental results, but it breaks down when the time dependence of the electroluminescence is examined. The most probable factor which causes the failure seems to be the assumption of a single-exponential isothermal thermoluminescence decay. Nevertheless, the most important conclusion that the electroluminescence of pre-irradiated polymer originates from the same population of trapped electrons as does the isothermal luminescence in the absence of an applied field continues to be valid.

3.8 Radiothermoluminescence of Organic and Inorganic Substances: Similarities and Dissimilarities

According to the current picture, when ionizing radiation is absorbed in a inorganic crystal, some electrons are displaced from the valance band or from impurity luminescence centers into the conduction band, in which state they may diffuse through the crystal [73]. An electron may remain mobile until it comes into a region where an electron is missing from the filled band (a positive hole) or becomes trapped at a position of relative stability, e.g., at a lattice imperfection. At a sufficiently low temperature, the electron may become localized by polarizing its surroundings, i.e., displacing the surrounding ions, to give a self-trapped electron, as predicted by Landau [74]. For each type of trap, there is a characteristic activation energy for release of the electron. As an irradiated crystal is heated, electrons are released from their traps by thermal activation and migrate through the crystal until they again become trapped or recombine either with positive holes trapped at different sites or with luminescence centers. The recombination of electrons with positive holes or with luminescence centers may result in the emission of photons, in which case the sample is said to exhibit thermoluminescence. Therefore, when light emission is measured during heating of an irradiated crystal, several temperatures of maximum emission are observed, each related to the activation energy for release of electrons from traps of some specific configuration. Exposure to light of appropriate wavelength also may cause release of electrons from traps and thus give rise to light emission. The behavior of holes is parallel to that of excited electrons; they may diffuse through the crystal and may become trapped at positive ion vacancies. Their release also may lead to light emission when the hole subsequently meets an excited electron. In fact, mainly electrons or positive holes will be released on warming, depending on the relative energy gaps between, on the one hand, the electron traps and the bottom of the conduction band and, on the other hand, the positive hole traps and the top of the valence band [75].

It is doubtful if exactly the same model can be applied to organic materials. The four important concepts embodied in this model are those of valence band, conduction band, electron trap, and recombination center or positive hole. If full formal analogy with inorganic materials is sought, then one can cite the formation of a valence band some 10 eV wide in paraffin polymers as deduced in a semiquantitative fashion by McCubbin and Gurney [76]. Presumably, the method can be extended to other polymer types, since the electronic interaction between neighboring repeat units and between neighboring chains in polymers is relatively small. Anyway, the lack of a well-established valence band concept is not a serious failing, since all that is required is a ''source'' of electrons which can be removed from bound states by ionizing radiation. Almost any form of molecular orbital will fulfill this requirement. The next consideration is the ejection of electrons either into a conduction band, where they are captured into electron traps, or directly into the electron traps. The direct-capture alternative means that the probable absence of a conduction band from polymers is of no consequence; one may consider folds in the chains, configurational imperfections in crystalline regions, or specific chemical groups as the electron traps. Finally, while the concept of a positive hole is closely related to that of valence band, alkyl radicals, additive residues, or the sites from which the electrons were originally ejected could act as efficient recombination centers. In addition, emission depended on the untrapping of both electrons, and positive holes should follow second-order kinetics, which is not the case with organic materials. Thus the absence of the description of the movement of positive holes in the model for charge liberation in organic substances seems not to be of importance. Therefore, the essentials of the framework on which most models for thermoluminescence in inorganic materials depend exist in organics.

In inorganic substances, the liberation of charges from the traps is purely thermal. According to Urbach [77], the depth of the trap E and the temperature T at which the charge may leave the trap are related as follows:

$$E \text{ (eV)} = T \text{ (K)}/500 \tag{3.56}$$

From Eq. (3.56) it follows that charges may be thermally liberated at temperatures under 250 K only from traps of depth not exceeding 0.5 eV. The energy of light quanta capable of removing electrons from such traps does not in any case exceed 1 eV (in ionic crystals and glasses, optical liberation of a trapped electron may need higher energy than thermal liberation by a factor of 1.2–2 [35]). However, during the radiolysis of organic substances, most of the electrons are stabilized mainly at deep levels and may be liberated from them only by the action of light quanta with energies of 2–3 eV [35]. Using Urbach's formula, one can estimate that the thermal liberation of electrons from such deep traps begins only at temperatures of 500–800 K, which exceed the melting point of most organic substances. Therefore, the recombination of ions on heating of an irradiated organic substance is governed primarily by the onset of molecular mobility and migration of the captured centers themselves, not by thermal liberation.

References

1. Magat, M.: J. Chimie Phys. *63*, 142 (1966)
2. Williams, F.: The Radiation Chemistry of Macromolecules, vol. 1 (ed. Dole, M.). Academic Press, New York, 1972
3. Keyser, R.M., Tsuji, K., Williams, F.: The Radiation Chemistry of Macromolecules, vol. 1 (ed. Dole, M.). Academic Press, New York, 1972
4. Nikolskii, V.G.: High Energy Chem. *2*, 233 (1968)
5. Charlesby, A.: Proceedings of the Fourth Symposium on Radiation Chemistry (ed. Hedvig, P., Schiller R.). Hungarian Academy of Science, Budapest, 1977
6. Medlin, W.L.: J. Chem. Phys. *38*, 1132 (1963)
7. Batsanov, S.S., Korobeinikova, V.N., Kazakov, V.P., Kobetz, L. I.: Optika i Spectroskopiya *30*, 484 (1971)
8. Partridge, R.H.: Polymer *23*, 1461 (1982)
9. Korobeinikova, V.N., Kazakov, V.P., Minsker, K.S., Soldaeva, N.P.: High Energy Chem. *6*, 326 (1972)
10. Charlesby, A., Partridge, R.H.: Proc. R. Soc. *A283*, 329 (1965)

11. Partridge, R.H.: The Radiation Chemistry of Macromolecules, vol. 1 (ed. Dole, M.). Academic Press, New York, 1972
12. Lewis, G.N., Kasha, M.: J. Am. Chem. Soc. *66*, 2100 (1944)
13. Charlesby, A., Partridge, R.H.: Proc. R. Soc. *A271*, 188 (1963)
14. Alfimov, M.V., Nikolskii, V.G., Buben, N.Ya.: Kinet. Catal. *5*, 238 (1964)
15. Keyser, R.M., Williams, F.: J. Phys. Chem. *73*, 1623 (1969)
16. Keyser, R.M.: ESR Study of Trapped Electrons in Gamma-Irradiated Hydrocarbon Polymers. Ph.D. dissertation, Univ. of Tennessee, Knoxville, Tenn., 1970
17. Dole, M., Bodily, D.M.: Adv. Chem. Ser. *66*, 31 (1967)
18. Ekstrom, A., Suenram, R., Willard, J.E.: J. Phys. Chem. *74*, 1888 (1970)
19. Bohm, G.G.A., Lucas, K.R.: Adv. Chem. Ser. *174*, 227 (1979)
20. Boustead, I., Charlesby, A.: Proc. R. Soc. *A315*, 271 (1970)
21. Partridge, R.H.: The Radiation Chemistry of Macromolecules, vol. 1 (ed. Dole, M.). Academic Press, New York, 1972
22. Boustead, I., Charlesby, A.: Eur. Polym. J. *3*, 459 (1967)
23. Altshuller, A.P.: Anal. Chem. *37*, 11R (1965)
24. Lifshits, I.M., Geguzin, Ya.E.: Sov. Phys. Solid State *7*, 44 (1965)
25. Burton, J.J.: Phys Rev. *177*, 1346 (1969)
26. Allen, N.S., Homer, J., McKellar, J.F.: J. Appl. Polym. Sci. *21*, 3147 (1977)
27. Yoshiaki Takai, Koji Mori, Teruyoshi Mizutani, et al.: J. Polym. Sci., Polym. Phys. Ed. *16*, 1861 (1978)
28. Charlesby, A., Partridge, R.H.: Proc. R. Soc. *A283*, 312 (1965)
29. Perekupka, A.G., Aulov, V.A.: Int. Polym. Sci. Technol. *8*, T/50 (1981)
30. Boustead, I.: Eur. Polym. J. *6*, 731 (1970)
31. McClain, W.M., Albrecht, A.C.: J. Chem. Phys. *43*, 465 (1965)
32. Brocklehurst, B.: Nature *221*, 921 (1969)
33. Frankevich, E.L.: Uspekhi Khimii *35*, 1161 (1966)
34. Yasuo Suzuoki, Teruyoshi Mizutani, Yoshiaki Takai, et al.: J. Appl. Phys. Jpn *16*, 1929 (1977)
35. Nikolskii, V.G., Tochin, V.A., Buben, N.Ya.: Sov. Phys. Solid State *5*, 1636 (1964)
36. Randall, J.T., Wilkins, M.H.F.: Proc. R. Soc. *A184*, 366 (1945)
37. Partridge, R.H.: J. Polym. Sci. [A] *3*, 2817 (1965)
38. Nikolskii, V.G., Tochin, V.A.: Adv. Radiat. Res. Phys. Chem. *2*, 535 (1973)
39. Fleming, R.J.: J. Polym. Sci. [A2] *6*, 1283 (1968)
40. Pender, L.F., Fleming, R.J.: J. Phys. [C] *10*, 1571 (1977)
41. Chen, R.J.: Mater. Sci. *11*, 1521 (1976)
42. McKeever, S.W.S.: Phys. Stat. Sol. (a) *62*, 331 (1980)
43. Williams, M.L., Landel, R.F., Ferry, J.D.: J. Am. Chem. Soc. *77*, 3701 (1955)
44. Knappe, W., Voigt, G., Zyball, A.: Coll. Polym. Sci. *252*, 673 (1974)
45. Bohm, G.G.A.: J. Polym. Sci. Polym. Phys. Ed. *14*, 437 (1976)
46. Bagdasaryan, Kh.S., Milyutinskaya, R.I., Kovalev, Yu.V.: High Energy Chem. *1*, 108 (1967)
47. Charlesby, A., Owen, G.P.: Int. J. Radiat. Phys. Chem. *8*, 343 (1976)
48. Randall, J.T., Wilkins, M.H.F.: Proc. R. Soc. *A184*, 390 (1945)
49. Mikhailov, A.I.: Proc. Acad. Sci. USSR *197*, 223 (1971)
50. Aulov, V.A.: Proc. Acad. Sci. USSR Phys. Chem. *254*, 812 (1980)
51. Partridge, R.H.: Polymer *23*, 1461 (1982)
52. Wintle, H.J.: Polymer *15*, 425 (1974)
53. Boustead, I.: Proc. R. Soc. *A319*, 237 (1970)
54. Boustead, I.: Proc. R. Soc. *A318*, 459 (1970)

55. Charlesby, A., Partridge, R.H.: Proc. R. Soc. *A271*, 170 (1963)
56. Nikolskii, V.G., Chkheidze, I.I., Buben, N.Ya.: Kinet. Catal. *5*, 69 (1964)
57. Meggitt, G.C., Charlesby, A.: Radiat. Phys. Chem. *13*, 45 (1979)
58. Brocklehurst, B., Evans, M., Stevenson, J.: Radiat. Phys. Chem. *15*, 361 (1980)
59. Aulov, V.A., Perekupka, A.G.: Vysokomol. Soedin. *B20*, 430 (1978)
60. Partridge, R.H.: Ph.D. Thesis, Univ. of London, 1964
61. Nakai, Y., Matsuda, K.: Jpn. J. Appl. Phys. *4*, 264 (1965)
62. Aulov, V.A., Sukhov, F.F., Slovokhotova, N.A., Chernyak, I.V.: High Energy Chem. *3*, 407 (1969)
63. Deroulede, A., Kieffer, F., Magat, M.: Israel Chem. J. *1*, 509 (1963)
64. Burton, M., Dillon, M., Rein, R.: J. Chem. Phys. *41*, 2248 (1964)
65. Curie, D.: Luminescence in Crystals. Wiley, New York, 1963
66. Somersall, A.C., Dan, E., Guillet, J.E.: Macromolecules *1*, 233 (1974)
67. Brocklehurst, B., Porter, G., Yates, J.M.: J. Phys.. Chem. *68*, 203 (1964)
68. Garlick, G.F.J., Gibson, A.F.: Proc. Phys.. Soc. *60*, 574 (1948)
69. Boustead, I., Charlesby, A.: Proc. R. Soc. *A315*, 419 (1970)
70. Ivey, H.F.: Electroluminescence and Related Effects. Academic Press, New York, 1963
71. Blake, A.E., Charlesby, A., Randle, K.J.: J. Phys. [D] *7*, 759 (1974)
72. Blake, A.E., Randle, K.J.: J. Phys. [D] *10*, 759 (1977)
73. Ghormley, J.A., Levy, H.A.: J. Phys. Chem. *56*, 548 (1952)
74. Landau, L.: Physik Z. Sowjetunion *3*, 664 (1933)
75. Fleming, R.J.: J. Polym. Sci. [A2] *6*, 1283 (1968)
76. McCubbin, W.L., Gurney, I.D.C.: J. Chem. Phys. *43*, 983 (1965)
77. Urbach, F.: Wiener Ber. *139*, 363 (1930)
78. Tochin, V.A., Shlyakhov, R.A., Sapozhnikov, D.N.: Polym. Sci. USSR *17*, 2934 (1975)
79. Tochin, V.A., Nikolskii, V.G.: High Energy Chem. *3*, 256 (1969)
80. Perekupka, A.G., Aulov, V.A.: Proc. Acad. Sci. USSR Phys. Chem. *247*, 712 (1979)
81. Perekupka, A.G., Aulov, V.A.: Vysokomol. Soed. [B] *22*, 578 (1980)
82. Boustead, I., George, T.J.: J. Polym. Sci. Polym. Phys. Ed. *10*, 2101 (1972)
83. Osawa, Z., Kuroda, H.: J. Polym. Sci. Polym. Lett. Ed. *20*, 577 (1982)
84. Osawa, Z., Kuroda, H., Kobayashi, Y.: J Appl. Polym. Sci. *29*, 2843 (1984)
85. Mckeever, S.W.S.: Thermoluminescence of Solids. Cambridge Univ. Press, New York, 1985
86. Ershov, B.G., Kieffer, F.: Chem. Phys. Lett. *25*, 576 (1974)

Chapter 4

Molecular Mobility and Transitions in Polymers

Polymers exhibit a variety of changes in state that drastically alter their physical and mechanical properties. The principal determinants in these transitions are temperature, external stress, and the time scale or rate of the experiment used to measure the transition under consideration. The temperatures at which these transitions occur for any particular time scale represent important material parameters.

If A represents a numerical value of some kind of polymer property (e.g., density, modulus, transparancy, etc.), then various cases of the temperature dependence of this property can be schematically represented as shown on Fig. 4.1. Obviously, any of the particular cases may take place at different temperature intervals. Case a is virtually equivalent to the case b if, instead of the factor A, its reciprocal value $1/A$ is considered (e.g., specific volume instead of density). Case a also can be transformed to case c if factor A is replaced by its derivative dA/dT. Then the deflection point on Fig. 4.1a corresponds to the maximum on Fig. 4.1c. Formally, it is easy to visualize the situation where not first, but following derivatives of A exhibit maxima. In the most common sense, a *transition* can be defined as the existence of a maximum of a certain property or its derivative at a certain temperature. The maximum can be very pronounced (up to discontinuity), or it may be spread out, covering a large temperature interval. Polymers, as a rule, exhibit very diffuse transitions. From the thermodynamic point of view, *transition temperature* is the temperature at which termodynamic potential derivatives undergo discontinuity. Discontinuity of first derivatives corresponds to first-order transitions, whereas discontinuity of second derivatives complies with second-order transitions.

Transitions are studied by conducting experiments in accordance with the scheme in Fig. 4.1. Usually, polymers exhibit multiple transitions which can be divided into two major groups: *phase transitions* and *relaxation transitions*. First are transitions from one phase state to another. They possess all the peculiarities of equilibrium thermodynamic transitions. Second are transitions that take place without changes in state. They are not thermodynamic, but rather are kinetic transitions which are gov-

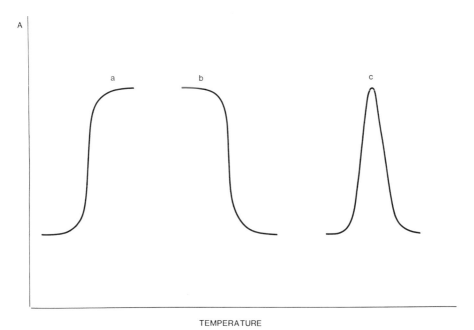

A

TEMPERATURE

Fig. 4.1. Schematic illustration of the changes in polymer properties with temperature in the transition region. After [31].

erned by relaxation processes in polymers. However, state transitions are affected by polymer relaxation, whereas relaxation transitions are accompanied by certain structural changes. Thus the classification of transitions is often not trivial, because kinetic and thermodynamic behavior are interrelated. Transitions in polymers are connected to the structure and reflect structural peculiarities. This indicates the importance of the study of the transitions in an attempt to establish structure–property relationships.

4.1 Phase Transitions

When a low-molecular-weight crystalline solid melts or when a liquid boils, changes in volume and enthalpy, as well as other thermodynamic properties, take place at constant temperature. Such changes are true thermodynamic changes of state. In this regard, there is no difference between polymers and low-molecular-weight substances. Polymers, of course, do not boil, but crystalline polymers undergo a transition similar to the melting transition of low-molecular-weight crystalline solids. There are, however, a number of important differences. For one, polymers melt or fuse over a temperature range, generally on the order of 2 to 10 degrees. The *equilibrium crystalline melting point* for polymers is defined as that temperature at which the last of the crystallites melt. Another important difference is that the actual value of T_m is subject to a strong hysteresis effect. That is, T_m depends on the melt history of the polymer as reflected in the percent crystallinity and the size distribution of the crystallites. Third, while low-molecular-weight materials become liquids on melting, polymers become very viscous (actually viscoelastic) fluids. These differences arise because the lengths of the polymer chains result in long-range interactions and physical entanglements. Self-diffusion and the flow of entire molecules in the polymer melt are thus restricted and produce the observed behavior. Segments of the chain, however, may be quite mobile, and considerable molecular flow can take place if the temperature or external stress is sufficiently high.

Both low- and high-molecular-weight organic solids may exhibit polymorphism; i.e., the substances are capable of forming three-dimensional arrays of quite different structure. Although one structure may be the more stable, the so-called metastable structure may still have great stability and above or below some critical temperature may be essentially permanent. Crystalline phase changes of this type are readily encountered in the paraffins as well as in some polymers. They are first-order crystal-crystal transitions involving sudden changes in volume, enthalpy, and entropy, and they do have an effect on the relaxation behavior of the solid.

State transitions can be considered from two points of view: thermodynamic and statistical. The first of these approaches is developed quite well. It considers equilibrium transitions in thermodynamic terms independently of a material's internal structure. The second approach deals with a model that more or less completely describes the real structure of a material. This is especially difficult in the case of polymers because of complications in their structure, and this approach is developed to much lesser extent than the first.

For first-order state transitions, transition temperature depends on the temperature changes in the chemical potentials of each phase. In Fig. 4.2, $\mu_1(T)$ and $\mu_2(T)$ represent the changes in chemical potential of phases 1 and 2, respectively. Each of phases is stable in a certain temperature region (*solid lines*) and becomes unstable at the transition temperature. At this temperature, $\mu_1 = \mu_2$. It is obvious that a phase transition takes place when the temperatures and pressures (in the absence of external fields) of both phases are equal. Because phase transition comprises an equilibrium process, isobar-isothermal potential (Gibbs free energy) at the transition temperature does not change:

$$dG = dH - T_0\, dS = 0 \quad \text{or} \quad dH = T_0 dS \qquad (4.1)$$

Since transition takes place at constant temperature, Eq. (4.1) can be presented in integral form:

$$T_0 = \Delta H / \Delta S \qquad (4.2)$$

where T_0 is transition temperature, and ΔS and ΔH are changes in entropy and enthalpy accompanying the transition.

In accordance with the thermodynamic definition of first-order transitions, first derivatives of the thermodynamic potential, namely, entropy and volume, that is,

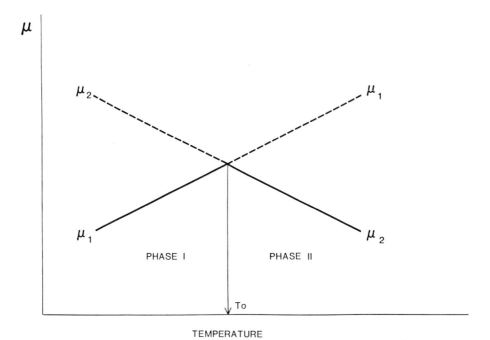

Fig. 4.2. Changes in the chemical potentials of two phases in the transition region. After [31].

$$\left(\frac{\partial G}{\partial T}\right)_p = -S \quad \text{and} \quad \left(\frac{\partial G}{\partial p}\right)_T = V$$

undergo discontinuity at the transition temperature.

First-order transitions include melting, crystallization, and phase transitions from one crystallographic modification to another.

4.2 Relaxation Transitions

Relaxation is defined as the transition from nonequilibrium to equilibrium state. In the absence of an external field, this transition entirely depends on the thermal motion of structural units which comprise the material. All factors which influence the mobility also influence relaxation. Temperature defines the kinetic energy of structural elements and the rate of their transition from one state to another. When temperature is raised, the mobility of structural elements and the rate of their transition from nonequilibrium to equilibrium conditions increase; i.e., relaxation processes proceed faster. The mobility of structural elements also depends on the energy of intermolecular interactions. If the system consists of nonpolar elements possessing small interaction energies, the equilibrium will be set faster than in a polar system with strong intermolecular interactions. If structural elements are small, they move faster than bulky and branched groups. Such systems will reach equilibrium more readily.

Thus high temperature, small size of structural elements, and small interaction energies promote the establishment of equilibrium in a system and, in turn, promote relaxation.

For a simple relaxation system (small structural elements and low interaction energies), changes provided by the thermal motion in an internal parameter p are proportional to the value of this parameter:

$$dp/dt = p/\tau \tag{4.3}$$

where τ is a coefficient of proportionality. After integration (from 0 to t and x_0 to x),

$$\ln (x/x_0) = - t/\tau \tag{4.4}$$

where x_0 is the initial value of the parameter of interest. The physical sense of the coefficient τ follows from the condition $t = \tau$. Then $\ln (x/x_0) = - 1$ and $x = x_0 e^{-1}$. Thus τ can be defined as the time during which the initial value of x decreases e times. τ is called the *relaxation time* and is a more convenient characteristic of the system than the total time needed for equilibrium.

Various systems are characterized by different values of τ. Low-molecular-weight fluids are characterized by a τ of 10^{-8}–10^{-10} sec. In polymers, the elementary chain units have dimensions similar to those of the molecules of low-molecular-weight substances. However, macromolecules are much longer, and their mobility is essentially smaller than the mobility of small molecules. The mobility of monomer units is also more restricted than the mobility of the same units not connected in chains by chemical links. The time for dislocation of macromolecular segments in elastomeres is of the order of 10^{-4}–10^{-6} sec. For macromolecules, relaxation time can be days or months.

Relaxation processes in polymers are considered as transitions because in their temperature intervals, material properties change more or less drastically. The relaxation transition divides two conditions of a material: one where τ is essentially larger than the time of observation for which relaxation can be neglected, and the other where τ is much smaller than the time of observation. In the latter case, the material is unchanged.

The common feature of relaxation transitions and thermodynamic transitions is that in both cases the relaxation time sharply changes in a relatively narrow temperature interval in comparison with the external time scale. However, differently from thermodynamic transitions, relaxation transitions are not associated with qualitative changes in the material's structure and are not connected with the equilibrium of two states of the material. The principal difference between thermodynamic transitions and relaxation transitions is that the first are independent of time scale and are equilibrium transitions and the second depend entirely on the time factor and are nonequilibrium processes.

The basic evidence proving the kinetic nature of relaxation transitions is that the transition temperature depends on a kinetic factor, namely, the cooling rate. However, experimental verification of this concept is not always simple, because relaxation time changes with temperature exponentially and small changes in temperature are associated with large changes in relaxation time, and vice versa. Thus, in order to observe small variations in the position of the relaxation transition (10–15°C), the cooling rate must be changed by several orders of magnitude.

In some cases it is suitable to distinquish between structural and mechanical relaxation. The first of these is associated with cooling, whereas the second takes place under dynamic action. For *structural relaxation,* the short order is "frozen in" and is connected with changes in internal time scale, i.e., the relaxation time of the system. *Mechanical relaxation* is realized when a polymer is deformed at isothermal conditions and depends on the external time scale, i.e., the time of external influence. In both cases, however, the kinetic nature of relaxation remains the same.

4.2.1 Nomenclature

Relaxation transitions are often labeled with Greek latters α, β, γ, etc. According to this nomenclature, the α transition corresponds to the relaxation observed at the highest temperature (at a given frequency) or the lowest frequency (at a given temperature). The β and γ symbols then apply to the other relaxation regions in order of decreasing temperature or increasing frequency.

For polymers in the amorphous state, a relaxation region associated with the glass transition is usually labeled α and is referred to as a *primary* or *glass-rubber relaxation.* This is by far the most pronounced relaxation. In addition to the glass-rubber relaxation, amorphous polymers usually exhibit at least one secondary relaxation region. This region (β, γ, or δ relaxations) results from motions within the polymer in the glasslike state. In this state, the main chains are effectively frozen in, so that these relaxations cannot be due to large-scale rearrangements of the main polymer chain.

The α, β, γ nomenclature is also used in labeling relaxations in crystalline polymers. The symbols do not imply in the case of crystalline polymers the same molecular mechanisms with which they are often associated in amorphous polymers [1]. The α label is often given to the region of relaxation associated with crystalline regions and observed very close to the melting point or some 50–100°C below the melting point of a crystalline polymer. In addition to the relaxation mechanisms occurring in the crystalline phase, crystalline polymers often show two transitions associated with the disordered or amorphous regions. The high-temperature peaks are attributed to movements of relatively large sections of chain in the amorphous regions, and these peaks are often clearly related to the glass transition. The lower-temperature amorphous peaks, which are labeled γ, are thought to involve limited motions of relatively short chain segments.

The confusing situation where glass-rubber relaxations in amorphous and crystalline polymers are labeled differently (α in amorphous and β in crystalline) reflects the difficulties that exist in understanding the nature of relaxation transitions, especially in crystalline polymers, and is to a certain extent a confession of ignorance of the precise molecular mechanisms involved.

4.2.2 Glass Transition

The glass transition T_g is one of the most important transitions in polymers. From a molecular point of view, it has been widely accepted for many years that glass-rubber relaxation results from large-scale conformational rearrangements of the polymer chain backbone. These rearrangements involve cooperative thermal motions of individual chain segments. Amorphous polymers exhibit a change from rigid, glasslike behavior below T_g to soft, rubbery behavior as the temperature is raised above T_g. The variety of changes that materials undergo at T_g predetermined repeated attempts to connect this phenomenon with internal material parameters. The most widely used parameter of this kind is the so-called free volume. The *free volume* can be defined as the difference between the real volume of a system at a certain temperature and the hypothetical volume which this system would have if its molecules were packed in a perfect crystalline lattice. The free volume characterizes the packing density of molecules and represents the population of holes having dimensions of atoms or small atomic groups. For each temperature, there is an equilibrium hole density, i.e., an equilibrium volume, and it contracts with temperature mainly because of a decrease in the density of the holes. In a liquidlike system, the local free volume is continually being redistributed throughout the medium, the redistribution occurring simultaneously with the random thermal motions of the molecules. The basic idea underlying the free-volume approach to relaxation phenomena is that molecular mobility at any temperature is dependent on the available free volume at this temperature. If the time necessary for molecular rearrangements is larger than the relaxation time, molecular rearrangements take place. As the temperature decreases, the free volume decreases and relaxation time increases. When temperature is lowered to such an extent that the relaxation time becomes larger than the time of molecular rearrangements, the internal reordering takes place so slowly that the volume cannot contract at a rate sufficient to maintain equilibrium, so the decrease in free volume cannot proceed any further. The specific volume of the polymer changes abruptly in the glass-transition region, indicating change in the packing conditions. This is schematically illustrated in Fig. 4.3, where the shaded area represents free volume and the point where volume-temperature dependence changes its slope corresponds to T_g. Above T_g, the hole density increases with increasing temperature and decreases with decreasing temperature. Below T_g, the hole density does not change when the temperature is changed. The hole density appropriate to a temperature close to T_g is "frozen in." It has been established experimentally that all polymers have approximately the same fractional free volume at T_g, a condition first proposed by Fox and Flory [2].

Under practical experimental conditions, the polymeric glasses are not in thermodynamic equilibrium. There is a structural relaxation process (sometimes referred to as *enthalpy relaxation*) which represents the time-dependent transition from the nonequilibrium (vitroid) state to the equilibrium (vitreous) state. The rate of this transformation is very low when the temperature is much below T_g, and it becomes faster on approaching T_g. The situation is schematically illustrated in Fig. 4.4. The sample is supposed to have been stored at temperature T_0 for an infinitely long time; thus thermodynamic equilibrium has been reached. By increasing the temperature, the

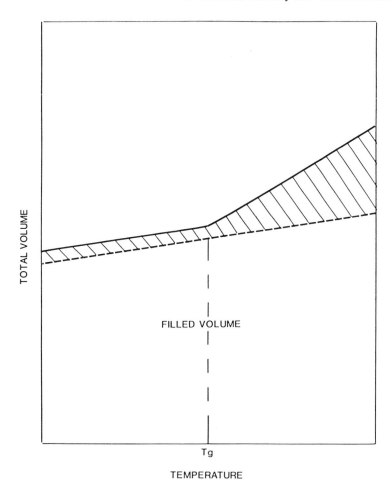

Fig. 4.3. Temperature dependence of the total volume, filled volume, and free volume (*shaded area*) for an amorphous polymer.

specific volume (and the free enthalpy) of the sample will exhibit a sharp break at a temperature T_g. The exact position and shape of this change depends on the rate of heating. Upon reaching point F, the material is transformed to the non-Newtonian liquid state. By cooling, the break is observed at a different temperature T_g^1, and a state referred to as a *vitroid* is reached (point 0^1) which has a higher free volume and a higher free enthalpy than the corresponding equilibrium state 0. By storing the sample at a temperature T_0, the free volume is decreased as the system tends to thermodynamic equilibrium. For a decrease in free volume, mobility must be decreased, but this decrease requires some free volume. Therefore, mobility cannot become 0 in a finite time. The state of zero mobility can be approached only asymptotically, and in practice, exact equilibrium is never reached. Correspondingly, measurements always start from a nonequilibrium state 0^{11}.

The kinetic theory of glass transition is based on the free-volume concept and

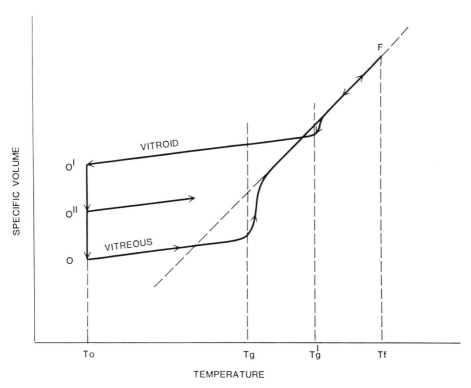

Fig. 4.4. Changes in the specific volume of a polymer by heating and cooling through the glass-transition temperature. After [3], by permission of Academic Press.

predicts the low temperature limit of T_g at infinitely slow cooling rates. The existence of this limit defines the difference between glass relaxation and other relaxation processes and also establishes a connection between polymer thermodynamic properties and relaxation parameters in the T_g region.

The other explanation of the low temperature limit of T_g is based on thermodynamic considerations. The entropy S of the chain molecules is proportional to the number of available conformations W and can be expressed as $S = k \ln W$. The flexibility of the chain molecules—their ability to chose among different configurations—is gradually lost with lower temperatures. With this loss in flexibility (and entropy), there occurs a greater and greater difficulty in finding ways to pack the molecules together on a lattice. The packing difficulty, inversely related to entropy, increases so severely as temperature is lowered that it becomes infinite, and the conformational entropy (and free volume) becomes zero at a finite temperature rather than at absolute zero. According to Gibbs and DiMarzio [4], this temperature, denoted by them as T_2, manifests the low temperature limit which can be reached when a substance is cooled down infinitely slowly. The scheme which illustrates this approach is shown in Fig. 4.5. Various dotted lines correspond to different cooling rates, and the large dots indicate different values of T_g. The entropy of the polymer in the glassy state

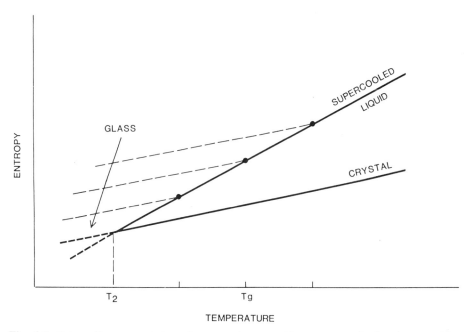

Fig. 4.5. Schematic representation of entropy changes with temperature in the glass-transition region. The arrow indicates the decrease in cooling rate. After [30], by permission of John Wiley & Sons, Inc.

depends on the kinetic of the process. The low-temperature limit of T_g (T_2) corresponds to the intersection of the temperature dependencies of the entropies of the liquid and the crystal. The equility of the entropies of the liquid and the glass indicates the attainment of the equilibrium structure.

The phenomenological aspects of glass transition are associated with the amorphous nature of polymers. Commonly speaking, the T_g of a semicrystalline polymer should be defined as the T_g of an amorphous polymer which is obtained from the crystalline material in question as a result of the destruction of its crystalline lattice. In practice, however, only in rare instances is it possible to suppress crystallization completely. With crystalline polymers that cannot be obtained in a completely amorphous state, there is always the problem of the proper destination of the glass-transition temperature.

4.2.3 Secondary Transitions

It has been known for many years that substances in the glassy state retain some degree of molecular mobility. These relaxations have been termed *secondary relaxations,* and they occur on a time scale many times shorter than that of the main relaxation responsible for the glass transition itself. The presence of such relaxations has generally been associated with the motion of a side group attached to a polymer chain or with a certain type of movement of the chain itself, and it has been explained in terms of hindered internal molecular modes of motion that remain active even when the molecule as a whole is frozen in place in a glassy matrix.

When a polymer is cooled and passes its glass transition T_g, the cooperative motion of large parts of the main polymer chain are frozen in. Since these motions require a considerable free volume, it is reasonable to suppose that below the glass transition, the available free volume is still large enough for the motion of smaller groups. Several types of motion possible below T_g have been proposed [5].

A. Local main-chain motion. This is a motion of main-chain segments smaller than those associated with the glass transition. Such a local main-chain motion is the only possible molecular mechanism causing a secondary transition in a polymer possessing no side groups that can move independently of the main chain. It is believed that type A motion is not sensitive to changes in free volume [6]. This is qualitatively understandable: At the glass transition, the cooperative motions of large parts of the chain require a considerable free volume; the motion of the much smaller groups that are responsible for secondary transitions requires only a small free volume; and after freezing in of the cooperative motion of the larger segments, this small volume always remains. Type A motion is rather general; it is found in polycarbonates, polysulphones, polyesters, and polyvinyl chloride, for example. From a practical point of view, local main-chain motion is of essential interest because the temperature at which it appears often marks the transition from brittle to tough behavior.

B. Side-group motion with some cooperation of the main chain (e.g., a partial rotation of the side group about the bond linking it to the main chain). The side group moves as one whole; its rotation need not be complete, and indeed, it is more likely to be a transition from one equilibrium position to another. The deformation of adjacent valence angles often forces the main chain to partake slightly in the side-group motion. A typical example of type B motion is found in polymethacrylates.

C. Internal motion within a side group itself, without interference by the main chain.

D. Motion of, or taking place within, a small molecule dissolved in the polymer. This can be motion within a plasticizer molecule. Small molecules dissolved in polymers sometimes associate themselves with side groups, particularly if both are polar. When association takes place and there is a combined motion, we get a type C instead of a type D motion.

4.2.4 Activation Energy of Relaxation

Activation energies such as must be provided, for example, in the conformation changes of the chain molecule to overcome the rotation potential are not considered in the free-volume theory. However, since relaxation time is a measure of the time required to approach equilibrium and in the actual polymer is thus inversely proportional to the rate of segmental diffusion, it is expected that τ will decrease with increasing temperature.

In practice, usually the value reciprocal to the relaxation time, namely, the relaxation frequency ($f = 1/\tau$), is measured. It is assumed that, in accordance with the Arrhenius equation, f varies exponentially with temperature:

$$f = f_0 e^{-\Delta G/RT} = f_0 e^{-\Delta H/RT} e^{\Delta S/R}$$

(4.5)

where ΔG, ΔH, and ΔS are, respectively, the activation energy, activation enthalpy, and activation entropy of the corresponding molecular motion, R is the gas constant, and T is the absolute temperature. The pre-exponential factor f_0 reflects the frequency of molecular vibration and is on the order of 10^{13} sec^{-1}.

The increase in f—or the increase in segmental diffusion rate—with temperature is a result both of thermal expansion leading to a greater free volume and of the greater probability of segmental jumps over energy barriers because of the higher thermal energies.

Depending on their activation entropies, viscoelastic relaxations can be considered to be simple or complex [7]. *Simple relaxations* have activation entropies near zero and reflect the motion of small molecular fragments without much cooperative involvement. *Complex relaxations* have large positive activation entropies and involve cooperative motions among neighboring groups of molecules. Most of secondary relaxations are simple, whereas glass transition is always a complex relaxation.

For simple relaxation, $\Delta S = 0$ and the relaxation frequency is expressed as

$$f = f_0 e^{-\Delta G/RT} = f_0 e^{-\Delta H/RT} \tag{4.6}$$

where $f_0 = 10^{13}$ sec^{-1}. The origin of activation energy in this case is mainly intramolecular.

The activation energies and/or pre-exponential factors for glass transitions usually are essentially higher than for secondary relaxations, indicating that there is a high degree of complexity in the motion associated with the relaxation. This impression of complexity is confirmed by large activation entropies for many of these relaxations [8].

The slope of the straight line in the activation plot ($\ln f$ versus $1/T$) is often called an "apparent" activation energy, which presumably is different from the "true" activation energy due to a temperature dependence of the latter. However, it can be easily proved that, first, a linear relationship between $\ln f$ and $1/T$ is observed only if ΔH is constant or depends linearly on temperature and that, second, such linear temperature dependence does not change the slope of the activation line but only its intercept on the ordinate [9]. Furthermore, a complete thermodynamic description of any relaxation process has to take into account the Gibbs free-activation enthalpy leading to Eq. (4.5). From Eq. (4.5) it follows with regard to the temperature dependence of ΔG that

$$-R \frac{\partial \ln f}{\partial (1/T)} = \Delta G - T \frac{\partial \Delta G}{\partial T} = \Delta H \tag{4.7}$$

The right-hand term in Eq. (4.7) is equal to the activation enthalpy, according to general principles of thermodynamics. Equation (4.7) shows that the slope $\partial \ln f / \partial (1/T)$ always yields the activation energy. The nonlinear temperature dependence of ΔH leads necessarily to a curved activation diagram.

There is another effect, however, which has to be regarded if the activation plot is discussed. The occupation number of the activated states also depends on its number of normal coordinates. If more than one unit participates simultaneously in the jump, the right-hand side of the Eq. (4.5) has to be multiplied by the following [10]:

$$\sum_{r=0}^{n-1} \frac{1}{r!} \left(\frac{\Delta G}{RT}\right)^r$$

where n is the number of units within a cooperatively moving region. This element causes an increase in the activation entropy and leads to a shift in the activation curve to lower temperatures. Such a shift, in turn, can be reflected in abnormally high values of the "apparent" pre-exponential factor (f_0') if it is expressed as

$$f = f_0 e^{-\Delta H/RT} e^{\Delta S/R} = f_0' e^{-\Delta H/RT} \tag{4.8}$$

Brot [11] offered another explanation for the large frequency-factor values often observed in polymers. According to him, these large values occur as a result of decreases in the height of the potential barrier as the temperature rises and the pressure remains constant owing to thermal expansion of the lattice:

$$f_0' = f_0 \exp\left[\frac{-\Delta H(1 - \alpha T)}{RT}\right] \tag{4.9}$$

where α is the linear coefficient of thermal expansion. It can be readily seen that Eq. (4.9) is formally identical to Eq. (4.5) if $\alpha\Delta H$ is taken in the sense of activation entropy.

4.3   Various Methods of Transition Analysis in Polymers

Since the early days of the use of natural and synthetic polymers, their relaxation has been realized as essential in characterizing their behavior and in predicting their properties during long-term use. This attitude has stimulated extensive development in various branches of relaxation spectroscopy. Relaxation spectroscopy generally involves subjecting the material to an external field and observing its response as a function of the characteristic parameters of the field [12]. In conventional optical spectroscopy, an electric field of light interacts with the electronic structure of the atoms or molecules. In magnetic-resonance spectroscopy, the magnetic field of a radiowave or microwave generator interacts with the permanent magnetic moments of the nuclei (nuclear magnetic resonance) or with the permanent electronic magnetic moments of the atoms, ions, or molecules (electron-spin resonance). In microwave spectroscopy of gases, the external radiation field interacts with the permanent electric dipole moment of the molecules, inducing changes in their rotational-vibrational states. A similar electric dipole-dipole interaction is effective in dielectric spectroscopy.

The primary effects in different branches of spectroscopy are due to the energy eigenvalues of the structutal units which interact with the external field. Through

Table 4.1. Qualitative comparison of methods of measuring transitions

Method	Effective frequency	Sensitivity	Specificity toward groups	Resolution
Specific volume	Low	Good	Nil	Good
Specific heat	Low	Fair	Nil	Fair
Dynamic mechanical loss	Low	Very good	Fair	Good
	High ($<10^4$ cps)	Good	Fair	Fair
Dielectric loss	Low	Good	Very good for polar groups	Very good unless electrolytes are present
	High ($<10^9$ cps)	Good	Very good for polar groups	Fair
Nuclear magnetic resonance	High (10^5–10^8 cps)	Very good	Very good for hydrogen and fluorine atoms	Varies from excellent to fair
Creep, stress relaxation	Low	Very good	Nil except for chemical creep	Good

Source: Ref. 13, © John Wiley & Sons, with permission.

secondary effects, however, the motion of other units also can be detected. A well-known example of this is nuclear magnetic resonance, in which the primary effect is change of orientation of nuclear magnetic moments by the magnetic field of the external radiation, but this effect is perturbed in the neighborhood of the nucleus so much that various studies of electronic and molecular structure and of molecular motion are made possible.

All the kinds of spectroscopy mentioned so far involve the interaction of electromagnetic radiation with matter. A particular type, the most widely used in relaxation studies, is mechanical spectroscopy, which involves the interaction of a periodic mechanical force field with the material and observations of the response of the material as a function of field frequency. The mechanical force field might consist of low-frequency external loading, sound waves, or even shock waves propagating through the material.

Various experimental techniques differ in sensitivity, specificity, and frequency (Table 4.1). Ideally, different test methods and a range of temperature and frequency for at least one of the test methods will be desirable for complete elucidation of multiple relaxation patterns and the mechanisms from which these relaxation regions arise.

4.3.1 Mechanical Spectroscopy

Dynamic-mechanical spectroscopy evaluates the ability of materials to store and dissipate mechanical energy on deformation. If, for example, a material is deformed and then released, a portion of the stored deformation (strain) energy will be returned at a rate that is a fundamental property of the material. That is, the material goes into damped oscillation. For an ideal elastic material, the energy incorporated into the oscillation will be equal to that introduced by the deformation, with the frequency of the resultant oscillation being a function of the modulus (stiffness) of the material. Most real materials, however, do not exhibit ideal elastic behavior, but rather they exhibit a viscoelastic behavior in which a portion of the deformation energy is dissipated in other forms, such as heat. The greater this tendency for energy dissipation, the greater is the damping of the deformation-induced oscillation. Hence dynamic-mechanical techniques facilitate the measurement of both the elastic (modulus) and viscous (damping) properties of viscoelastic materials. The information obtained (especially the information provided by the damping) is of great importance, since such end-use properties as vibration dissipation, heat buildup, impact resistance, and noise abatement are all related to mechanical damping. These tests have also proven useful in evaluating transitions' in polymers.

The common feature of relaxation transitions, namely, their dependence on duration (or frequency) of the external field, can be illustrated by analyzing the temperature dependence of mechanical losses in polymers. Let us consider a so-called standard linear solid, in which a simple elastic element and a simple viscous element which are placed in parallel are combined in series with another simple elastic element. This model can be characterized by three constants, an elastic modulus G_1, an associated viscosity η_1, and a second elastic modulus G_0 (Fig. 4.6). Such a model roughly, but qualitatively correctly describes creep (with a characteristic retardation time $\tau_s = \eta_1/G_1$), stress relaxation (with a characteristic relaxation time $\tau_\alpha = \eta_1/G_0 + G_1$), and dynamic losses (with a characteristic time $\tau = \sqrt{\tau_\alpha \tau_s}$) in polymers.

If a sinusoidal stress of circular frequency ω is applied, the internal friction, or tangent of the loss angle, is given by

$$\tan \delta = \frac{G_0}{\sqrt{G_1(G_0+G_1)}} \frac{\omega\tau}{1+\omega^2\tau^2} \tag{4.10}$$

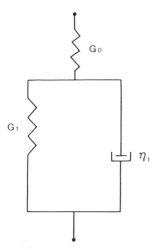

Fig. 4.6. A model of a standard linear solid.

From Eq. (4.10) it is seen that the internal friction rises from very low values at low frequencies to a maximum at a frequency for which $\omega\tau = 1$ and that it then falls again to low values at high frequencies. Furthermore, if the relaxation time τ varies exponentially with temperature, as it often does [that is, if $\tau = \tau_0 \exp(\Delta G/RT)$], then, on holding frequency constant and varying the temperature, similar behavior will be found; that is, once again the internal friction will pass through a maximum at a temperature for which $\omega\tau = 1$.

Thus there is a temperature at which the mechanical losses are maximum; at this temperature, the characteristic time of the external field ($1/\omega$) becomes equal to the internal time scale of the material (τ). The experimental values of ω may vary from fractions to millions of hertz. It is obvious that the τ value at which the condition $\omega\tau = 1$ is fulfilled depends on the chosen frequency. Therefore, the temperature of the maximum of mechanical losses is also frequency-dependent.

All these consequences are descriptive. Nevertheless, they show that the relaxation transition is accompanied by the maximum of the mechanical losses and that the temperature of this transition depends on the frequency (time) of the external field. In real polymers, relaxation is usually not restricted to a single process, but there is a set of more or less sharply divided processes which are characterized by their own τ and ΔG parameters. If the system is characterized by a number of discrete transitions which have essentially different τ values and thus significantly different positions on the temperature scale, the preceding analysis can be separately applied to each of these transitions. Experimentally, a series of relaxation processes with largely different characteristic times is reflected by the appearance of several discrete maxima on the mechanical losses–temperature curve.

The main complication arises when a polymer exhibits a discrete or infinite set of relaxation times. In this case, because of superposition of various relaxation processes, the temperature position of the loss maximum will depend on the combination of relaxation times. The character of the shift of the maximum along the temperature scale also will be changed, since different relaxation times may be characterized by different activation energies. The superposition of various relaxation processes may lead to the "smearing out" of a loss maximum or to the splitting of a maximum over several maxima when the frequency is changed.

Generally, dynamic-mechanical measurements for which both stress and the resulting strain are sinusoidal functions of time can be divided into four categories (which in some cases overlap) covering, in all, a frequency scale of about $0.1-10^8$ Hz. These four categories are (1) free-vibration methods, (2) forced vibrations in resonance, (3) non-resonance-forced vibration methods, and (4) wave-propagation methods.

Methods involving free vibrations cover the frequency range from $0.1-10$ Hz. The most commonly used apparatus in this group is the torsional pendulum. The specimen, usually with an additional mass attached, is given an initial displacement sufficient to cause it to vibrate. Mechanical losses are obtained by measurement of the decay in amplitude of vibration, the logarithmic decrement Δ being determined. At low values of damping, Δ is related to $\tan\delta$ by the following relationship:

$$\tan \delta \simeq \Delta/\pi$$

Forced-vibration measurements, in which a specimen is subjected to an impressed force varying sinusoidally, can be performed either with or without added mass and can be carried out at the resonance frequency of a free or loaded sample. If no additional mass is present, the total mass of the system is that of the sample. In this case, the resonant frequency is dependent on sample dimensions and density. The resonance vibration of nonloaded specimens has been carried out by the transverse vibration of reeds, thin strips and rods, and the bending, longitudinal, or torsional vibrations of rods. This general type of measurement covers a frequency range 10–10^5 Hz. The damping, or internal, friction Q for small values of damping can be related to the logarithmic decrement by the simple relation

$$Q^{-1} \simeq \Delta / \pi$$

Non-resonance-forced vibration methods can be carried out either with or without added mass. With these methods, a small specimen is subjected to alternating shear, and the tangent of the loss angle is obtained as a ratio of loss compliance and storage compliance. The frequency range attainable by these measurements is about 10^{-2}–10^4 Hz. The damping is obtained from the phase angle. Furthermore, if a non-resonance-forced method is employed, a larger frequency range can be measured than with resonance methods. It is also simpler experimentally to keep the frequency constant while varying the temperature. A well-known forced-vibration nonresonant instrument is the Rheovibron.

Wave-propagation methods are especially applicable at high frequencies (10^5–10^8 Hz). In this type of apparatus, one sends a pulse down a rod and measures the transit time of the pulse. The measure of damping, the attenuation of the pulse α, is given in units reciprocal centimeters. For small damping and the propagation of shear waves,

$$\tan \delta = \frac{\alpha \lambda}{\pi}$$

where λ is the wavelength of the propagating wave.

4.3.2 Dielectric Spectroscopy

Dieiectric spectroscopy is based on the response of a material to disturbances caused by an externally applied electric field. The results of alternating-field experiments can all be reduced to a statement of the behavior of the dielectric constant (ϵ') and the dissipation factor ($\tan \delta = \epsilon''/\epsilon'$) as a function of frequency and temperature.

The dielectric properties of nonpolar polymers approach the ideal behavior (ϵ' independent of frequency, $\epsilon''/\epsilon' = 0$). In contrast, for polar materials, the dielectric constant decreases with an increase in frequency and the dissipation factor increases and decreases in a cyclic manner. It is customary to plot the dielectric constant and its product with the dissipation factor, the loss index ϵ'', as a function of frequency. It should be noted that the absorption peaks (loss-index maxima) occur at the frequencies at which the dielectric constant is changing most rapidly. At high frequencies, the absorption peak is caused by polarization of dipole orientation; this is related to the "frictional" losses resulting from the displacement of molecular dipoles under the influence of the alternating field. Such dipoles may themselves be induced by the electric field and tend to align with it. In molecularly symmetrical materials it is impossible to induce a dipole in the pure material, so the material is considered nonpolar. However, in polar materials the molecular symmetry is disturbed and dipoles are polarized. At low frequencies, the accumulation of charges takes place at the interfaces between phases of the material which may have different dielectric constants. Such an effect is called *interfacial polarization.*

For nonpolar materials, the dielectric constant changes very little, if at all, with frequency and temperature. However, many nominally nonpolar materials contain traces of polar impurities. These impurities have relatively little effect on the dielectric constant, but they do increase the dissipation factor. For polar polymers, changes in ac characteristics with frequency are much more complex. At very high frequencies, crystalline barriers or the viscosity of a glassy polymer will prevent the dipoles from moving at all in the very rapidly reversing electric field. Consequently, the dipole structure will not influence the ac characteristics at very high frequencies. As the frequency decreases, the dipoles will move to a limiting extent, accompanied by energy loss. The loss will increase as the frequency decreases, and the dipoles can more and more follow the reversals of the electric field. Ultimately, the reversals become slow enough for the dipoles to encounter less resistance to moving, and loss again decreases. At the lowest frequencies, all the dipoles that will ever turn under the influence of the electric field will do so and the dielectric constant ϵ' will have its highest, or "static," value in so far as dipole contribution is concerned. At the same low frequency, the loss index ϵ'' will reach a minimum.

Temperature affects the dielectric relaxation time in a manner similar to mechanical relaxation. Thus it can be readily appreciated that the loss index at different frequencies will also pass through maxima as a function of temperature.

Most of the recording dielectric spectrometers are operated at a fixed frequency by recording the dielectric constant and the loss index as a function of temperature. Technically, it is much easier to measure dielectric parameters as a function of frequency than mechanical ones. It is a feature of the dielectric technique that measurements can be performed nearly continuously over the frequency range $10^{-4}-10^{10}$ Hz, and it is essential that as large a frequency range as possible should be covered, since dielectric relaxation curves for polymers are broad and very sensitive to temperature variation.

A special kind of dielectric spectroscopy, the so-called dielectric depolarization technique, has recently been developed in different laboratories. The technique is very simple. For studying relaxation transitions, the circuit method, which involves measurement of the current of a previously polarized sample in short circuit without external voltage, is most useful. Polarization is accomplished by application of a dc voltage above the highest-temperature transition to be studied. The current peaks obtained as a function of temperature can then be correlated to dipole relaxation [14].

4.3.3   Nuclear Magnetic Resonance Spectroscopy

In nuclear magnetic resonance spectroscopy, the interaction of the electronic or magnetic spin system with an external magnetic field is measured. The spin system is
polarized by a constant magnetic field, and magnetic dipole transitions between the
magnetic energy levels are induced by an additional sinusoidally oscillating magnetic
field polarized perpendicular to the polarizing field. It is useful to identify two distinct systems within the array of molecules. One system consists of the assembly of
molecules as a whole, called the *lattice*. The other system consists of the assembly
of nuclear spins, i.e., the *nuclear magnetic dipoles*. The response of the spin system
to the external magnetic field is conventionally characterized by two relaxation times:
the spin-lattice relaxation time T_1 and spin-spin relaxation time T_2. The first describes
the radiationless transition from an excited spin level to the ground state—the excess
energy being transferred to the surroundings (lattice). The second describes the mutual interaction between individual spins; it governs the decay of the magnetization
components perpendicular to the direction of the polarized field.

The two systems, the lattice and the nuclear spins, are closely coupled. Interaction
of the magnetic dipoles with the magnetic field at usual field strength is much stronger
than their interaction with anything else in their environment. The polarization of the
magnetic dipoles by the magnetic field is so dominant that their orientation with
respect to the applied field is little affected by molecular motion. If a system of nuclei
has been polarized by a magnetic field and the field is then turned off, the decay of
this polarization through the agency of molecular motion is found to occur very
slowly on the time scale of such motion and molecules must undergo many reorien-
tations before the total magnetic interaction is appreciably changed. Indeed, relaxation of the magnetic polarization may take seconds or even many minutes, whereas
the time scale of molecular rotation or translation in solids or liquids at room temperature is typically in the range 10^{-5}–10^{-10} sec or even shorter. This situation
should be contrasted with dielectric and mechanical relaxation. In a dielectric experiment, the permanent electric dipole moves directly with the molecule. When an
applied electric field is removed from a polar substance, the macroscopic electric
polarization decays at a rate comparable with the average time of molecular reorientation. Similarly, in a viscoelastic experiment, the removal of a stress is followed by
a relaxation of strain on a time scale comparable with that for molecular reorientation
or translation.

Because the spin system and the lattice are very weakly coupled, it is valid to
ascribe a temperature T_s to the spin system which may be different from that of the
lattice. At thermodynamic equilibrium, T_s equals the lattice temperature T_L, which is
the quantity conventionally measured. Irradiation at the resonant frequency in a magnetic field causes T_s to rise above T_L. The recovery of thermal equilibrium when
irradiation is ended may be thought of as a cooling down of the spin system. It
depends on the spin-lattice relaxation process and is characterized by a time constant
T_1.

In nuclear magnetic relaxation, the direct effect of molecular motion is to change
the interactions among the magnetic dipoles on a time scale characteristic of molec-

ular reorientation. The establishment of thermodynamic equilibrium between the spin system and the lattice is commonly observed to be a first-order rate process. The polarization of an excited spin system decays to equilibrium polarization as

$$P(t) = P_0(1 - e^{-t/T_1})$$
(4.11)

where T_1 is the time constant for the exponential approach toward equilibrium. $P(t)$ is proportional to the total magnetization of the sample in the direction of the polarizing field (longitudinal magnetization). The other relaxation time, T_2, also decays exponentially in good approximation. It describes the mutual interaction between individual spins and governs decay of the magnetization components perpendicular to the direction of the polarizing field.

Spin-lattice relaxation through molecular motion occurs most efficiently when the characteristic frequency of motion is comparable to the nuclear magnetic resonance frequency, typically 10^6–10^8 Hz. In experiments at fixed frequency and variable temperature, one ordinarily observes that T_1 passes through a minimum at some temperature (or perhaps a number of minima at different temperatures). Such a minimum marks a transition occurring at a frequency of motion comparable to the radiofrequency of the experiment. The temperature of the nuclear magnetic resonance transitions is found to depend on the nuclear magnetic resonance frequency, reflecting the temperature dependence of the spectrum of molecular motion. As with other kinds of relaxation measurements, the occurrence of multiple T_1 minima is ascribable to different types of motion that become predominant at the nuclear magnetic resonance frequency at different temperatures.

The spin-spin relaxation time T_2 as plotted against temperature shows no minimum. T_2 is constant over a range at low temperatures and increases at temperatures corresponding to extensive motion of whole-chain segments (T_g region).

In nuclear magnetic resonance spectroscopy, the temperature must be chosen as an independent variable. Since the molecular motion is indirectly detected through the interaction of the spin system with the different subsystems of molecular motion, it is difficult to obtain straightforward information about the molecular motion correlation process from the nuclear magnetic resonance data. Usually, the single-correlation-time approximation is applied. This is the average correlation time measured at temperatures corresponding to the megahertz range (for spin-lattice relaxation) which is useful for correlating nuclear magnetic resonance data with those measured by other methods.

4.3.4 Dilatometry and Calorimetry

In some respects, dilatometry and calorimetry are not as sensitive as other techniques, and only transitions which are highly cooperative (require the participation of many molecules simultaneously) and occur over a narrow temperature range can be detected. Measurements of specific volume (dilatometry) and specific heat (calorimetry) as a function of temperature permit the evaluation of such major transitions as glass transition and crystalline melting. However, there are no abrupt changes in these parameters from near 0 K to T_g similar to the mechanical or dielectric relaxations observed below T_g. The most probable reason for such "insensitivity" is that specific volume and specific heat are averaged over all molecular frequencies from very low to optical frequencies.

Nevertheless, the most important parameters of free-volume theory, such as specific volume, equilibrium and nonequilibrium free volumes, and expansion coefficients, are determined by dilatometry.

Calorimetric measurements are of considerable importance in the interpretation of relaxation spectra. They have a high discriminatory potential and can distinquish between different kinds of transitions independently. Specific heat exhibits no steps below T_g and therefore can be used as an unambiguous assignment of T_g. This is especially important in semicrystalline polymers, where multiple relaxations are observed.

4.4. Connection Between Mechanical Properties and Transitions in Polymers

Two kinetic processes of different physical natures may progress in the solid under load: deformation and fracture. The independence of deformation and fracture should be understood in the sense that they may occur independently and are processes of different physical nature (their rates are determined by different potential barriers, and their elementary events occur in different activation volumes). Actually, under strain experimental conditions, one of them may strongly affect the other, becoming even a controlling process. In this case, it is impossible to separate the parameters of these processes. If a load is applied to a specimen, the micro overstresses considerably exceed the average value. Further, two competitive processes are available: relaxation of overstresses and fracture. If owing to casual circumstances (low temperatures, rapid stress application) there is no time to reduce the local overstresses due to relaxation, the submicrocracks which are generated link and form a main crack which makes fracture inevitable. If the overstresses have somewhat relaxed, the fracture rate will sharply fall and the specimen either will not break at all or will break only after a long time. Competition between the processes of deformation (relaxation) and fracture must take place at all temperatures, and the deformational ability of polymers is strongly associated with the molecular mobility of their chains. The change in the character of molecular motion with temperature in many cases is accompanied by a change in the mechanism of fracture and defines the toughness (strength) temperature dependence of the material.

Low-frequency experiments at room temperature revealed that the difference between ductile and brittle behavior in polymers is directly linked to two activation energies: ΔG, the activation energy for glass transition, and U_0, the activation energy for mechanical fracture as extrapolated to 0 load (bond-rupture energy). It was found that when $\Delta G > U_0$, the polymer is brittle, and when $U_0 > \Delta G$, the polymer is ductile [15]. Such a relationship certainly indicates that segmental mobility dictates whether a polymer fails in a brittle or a ductile manner.

By assuming that the rate-determining step in polymer fracture involves secondary-chain motions that normally occur at low frequencies at the experimental temperature (below T_g), Sacher [16] showed that the time to break for a polymer in uniaxial tension τ can be represented as

$$\tau = \tau_0 \exp\left[(2\Delta H - \gamma\sigma)/RT\right] \tag{4.12}$$

with

$$\tau_0 = h/[kT \exp(2\Delta S/R)]$$

where ΔH and ΔS are the enthalpy and entropy of activation, respectively, R is the gas constant, $\gamma = \Delta V$ is the molar volume of activation for the fracture, σ is stress, T is the temperature, k is the Boltzman constant, and h is the Plank constant. The coefficient 2 appearing in the Eq. (4.12) reflects that both sides of an incipient crack contribute equally to its formation. Equation (4.12) is similar to the famous Gurkov's phenomenological equation of "time to fracture" and adequately describes not only the material behavior in uniaxial tension, but also such processes as polymer friction [17,18] and wear [19], the mechanical degradation of polymeric solids [20], and the time between stress application and the initiation of fiber drawing [21]. It may be that the mechanism based on molecular motion also applies to these cases.

There is much evidence that ultimate mechanical properties depend on those motions thought to contribute only to the linear viscoelastic region of polymers. The most widely known example of this kind is the impact strength of polymers, where the high strain rate of the impact precludes contributions from the nonlinear region.

By evaluation of a wide variety of polymer films at high strain rates, Sacher [22] was able to establish a direct relationship between toughness (the integrated stress-strain curve) and dissipation factor tan δ at 25°C, extrapolated to 5 kHz, a reasonable impact frequency (Fig. 4.7). Only one material among eight evaluated fell significantly off the straight line. The stress-strain curve of this particular material showed the reason for this: the material elongated some 200% with toughness at yield stress accounting for only 1.65% of total toughness. That is, plastic deformation contributed more than 98% of the toughness. Since the plastic contribution for this material far overshadows the elastic contribution and the dissipation factor is measured in the elastic region, correlation between the two is not expected. However, when the total toughness of this material was replaced by the toughness at yield stress, the (solid) point falls gratifyingly close to the line in Fig. 4.7. This finding correlates with the results of Retting [23] and Bauwens [24], who showed that the yield stress of polyvinyl chloride is related to the relative position and intensity of the test conditions with respect to the secondary transition, and the results of Broutman and Kobayashi [25], who observed that the fracture energy of polymethyl methacrylate was directly related to both the glass and the secondary transitions.

The correlation between the frequency dependence of fatique-crack propagation (FCP) in polymers and the frequency of secondary transition at the test temperature was noted by Manson et al. [26]. The sensitivity of FCP to frequency varied widely from polymer to polymer. The crack growth rates of polycarbonate and nylon-6,6 were little affected by changing the frequency from 0.3 to 10 Hz, whereas two- to

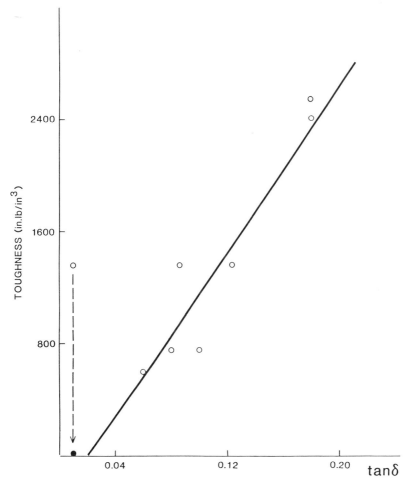

Fig. 4.7. Plot of toughness versus dissipation factor for a variety of polymer films. The solid point represents the toughness of Tedlar at the yield stress. After [22], © John Wiley & Sons, with permission.

threefold effects (the lower the crack growth rates, the higher was the frequency) were observed for several other polymers, such as polymethyl methacrylate and polystyrene. To characterize frequency response, a *frequency-sensitivity factor* (*FSF*) was defined as the multiple by which the FCP rate changes per decade change in frequency. Figure 4.8 shows the relationship between the FSF and the frequency of secondary relaxation at the test temperature and also indicates the range of frequencies used in the fatique tests. From the data presented in Fig. 4.8, it certainly seems likely that the greatest sensitivity of FCP frequency occurs when the test frequency is close to the frequency of the relaxation process. In all the insensitive materials, the frequency is far above the test range, whereas in the more sensitive materials, the relaxation frequency tends to be close to or within the test range.

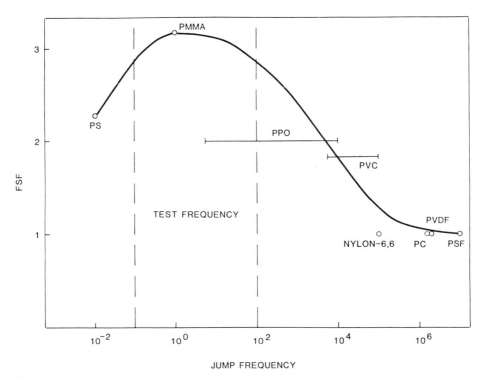

Fig. 4.8. Relationship between fatique-crack propagation frequency sensitivity and the room temperature jump frequency for several polymers: polystyrene (PS), polymethyl methacrylate (PMMA), polypropylene oxide (PPO), polyvinyl chloride (PVC), nylon-6,6, polyvinylidene fluoride (PVDF), polycarbonate (PC), and polysulphone (PSF). After [26], by permission of IPC Science and Technology Press, Ltd.

There is often a close correlation between loss peaks, as measured by dynamic-mechanical or electrical tests, and toughness. A particularly good example of such a correlation was found in polytetrafluoroethylene [27], in which three well-defined loss peaks appear to coincide with similar peaks in Izod impact strength (Fig. 4.9). The picture is far from clear, however, since, although they are good in some materials, the correlations do not occur in others, often because of the interaction of such phenomena as ductile-brittle transitions, notch sharpness effects, and environmental factors. If there is a trend in impact strength without a corresponding trend in material damping characteristics, this indicates that the cause is a change in the severity of structural defects or in some other factor which does not affect molecular mobility [28]. Conversely, if a trend in impact strength is closely correlated with a trend in dynamic-mechanical properties, this implies that the change in impact behavior is caused by a change in molecular or segmental mobility. In this connection, it is particularly important to bear in mind that an impact stress pulse is composed of a broad distribution of frequencies which extend to kilohertz values. If the damping occurs over the same distribution of frequencies generated by the impact, then the amplitude of the impact stress wave will be attenuated and the polymer will have high impact resistance. If the two frequency distributions do not have much overlap, there will be little attenuation and, all other things being equal, the polymer will have poorer impact resistance. The frequency, determined dielectrically or mechanically, will be higher the lower is the temperature of the $T<T_g$ relaxation at 1 Hz and the higher its energy of activation. According to the time-temperature superposition principle, measurements made at low temperature correspond to measurements made at high frequency, and vice versa. In the glassy state, where secondary transitions occur, the relation between measurements made as a function of absolute temperature T at a fixed frequency f_0 and measurements made at a fixed temperature T_0 as a function of frequency f is given by an Arrhenius-type equation:

$$\ln\,(f/f_0) = (\Delta G/R)\,(1/T - 1/T_0) \tag{4.13}$$

where ΔG is the activation energy of relaxation, and R is the gas constant.

A detailed study of the relation between impact resistance of *bis*-phenol A polycarbonate and its dynamic mechanical characteristics has been conducted by Hartmann and Lee [29]. This study showed that a secondary-transition damping peak observed centered at 170 K (at 1 Hz) is equivalent to a broad distribution of frequencies (at 293 K) from 1 Hz to 1 MHz, centered at about 100 Hz. On the other hand, an Izod impact test on this polymer at 293 K is equivalent to testing over the frequency range from 0–1000 Hz, the most important contributions (those with the biggest magnitude) being below 200 Hz. The overlapping of these two frequency

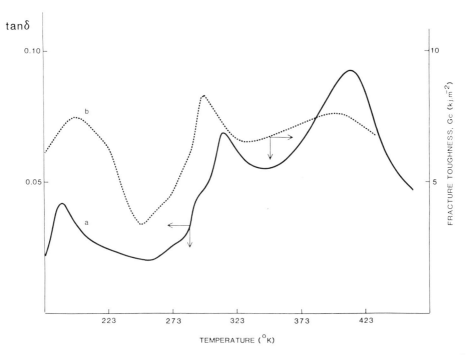

Fig. 4.9. Loss factor at 11 Hz (*a*) and impact fracture toughness at a strain rate of 72 sec^{-1} (*b*) versus temperature for commercial polytetrafluoroethylene. After [27], by permission of IPC Science and Technology Press, Ltd.

ranges represents an explicit, although qualitative, demonstration of the physical mechanism whereby a secondary transition occurring well below room temperature has determinant influence on polymer impact resistance.

References

1. McCrum, N.G., Read, B.E., Williams, G.: Anelastic and Dielectric Effects in Polymeric Solids. Wiley, London, 1967
2. Fox, T.G., Flory, P.J.: J. Appl. Phys. *21*, 581 (1950)
3. Hedvig, P.: The Radiation Chemistry of Macromolecules, vol. 1 (ed. Dole, M.). Academic Press, New York, 1972
4. Gibbs, J.H.: Modern Aspects of the Vitreous State, vol. 1, chap. 7. Butterworths, London, 1960
5. Heijboer, J.: Int. J. Polym. Mater. *6*, 11 (1977)
6. Bree, H.W., Heijboer, J., Struck, L.C.E., Tak, A.G.M.: J. Polym. Sci. Polym. Phys. Ed. *12*, 1857 (1974)
7. Starkweather, H.W.: Macromolecules *14*, 1277 (1981)
8. Starkweather, H.W.: Macromolecules *13*, 892 (1980)
9. Fisher, E.W., Goddar, H., Piesczek, W.: J. Polym. Sci. [C] *32*, 149 (1971)
10. Koppelmann, J.: Kolloid-Z. Z. Polym. *216*, 6 (1967)
11. Brot, C.: Disc. Faraday Soc. *48*, 213 (1969)
12. Hedvig, P.: J. Polym. Sci. Macromol. Rev. *15*, 375 (1980)
13. Boyer, R.F.: J. Polym. Sci. [C] *14*, 3 (1966)
14. Hedvig, P.: Dielectric Spectroscopy of Polymers. Halsted, New York, 1977
15. Aharoni, S.M.: J. Appl. Polym. Sci. *16*, 3275 (1972)
16. Sacher, E.: J. Macromol. Sci. Phys. *B15*, 171 (1978)
17. Schallamach, A.: Wear *1*, 384 (1957)
18. Moore, D.M., Geyer, W.: Wear *22*, 113 (1972)
19. Ratner, S.B., Lure, E.G., Radyukevich, O.V.: Sov. Plast. 6, 64 (1968)
20. Bueche, F.: J. Appl. Polym. Sci. *4*, 101 (1960)
21. Ender, D.H., Andrews, R.D.: J. Appl. Phys. *36*, 3057 (1965)
22. Sacher, E.: J. Appl. Polym. Sci *19*, 1421 (1975)
23. Retting, W.: Eur. Polym. J. *6*, 853, (1970)
24. Bauwens, J.C.: J. Polym. Sci. [C] *33*, 123 (1971)
25. Broutman, L.J., Kobayashi, T.: International Conference on Dynamic Crack Propagation. Lehigh Univ., Lehigh, Pa., 1972
26. Manson, J.A., Hertzberg, R.W., Kim, S.L., Skibo, M.: Polymer *16*, 850 (1975)
27. Kisbenyi, M., Birch, M.W., Hodgkinson, J.M., Williams, J.G.: Polymer *20*, 1289 (1979)
28. Vincent, P.I.: Polymer *15*, 111 (1974)
29. Hartmann, B., Lee, G.F.: J. Appl. Polym. Sci. *23*, 3639 (1979)
30. O'Reilly, J.M., Karasz, F.E.: J. Polym. Sci. [C] *14*, 49 (1966)
31. Adrianova G.P.: Physico-chemistry of polyolefins, Chimia, Moscow, 1974

Chapter 5

Radiothermoluminescence as a Method of Analysis of Transitions in Polymers

5.1 Radiothermoluminescence of Organic Substances and Molecular Motion

The general mechanism of radiothermoluminescence in both inorganic and organic materials consists of ionization within the material by the incident radiation, with trapping of some of the resulting ions if the temperature is low enough. Subsequent warming promotes recombination of the ions and thus formation of neutral excited molecules, which, in some cases, return to the ground state by radiative emission. However, the mechanism of electron release from the traps appears to be very different in ionic crystals from what it is in molecular solids. In ionic crystals, in fact, it is believed that the activation energy is imparted to the electron, permitting it to escape from a trap which continues to exist. By contrast, the activation energy in organic solids probably serves to destroy the trap and thus to liberate the negative charge.

Table 5.1 presents the positions of the thermoluminescence peaks for a certain number of organic compounds which are known to possess points (or regions) on the temperature scale at which either phase or relaxation changes occur [1–3]. It can be seen that for all these compounds, most thermoluminescence peaks correspond to the transition temperatures. This highly important observation was first made by Nikolskii and Buben [1] and later was generalized by Semenov [3], who related it to an analogous observation concerning the recombination of radicals. There is, nevertheless, some ambiguity in the results: A certain number of peaks appear at temperatures other than those of the known transitions. The question that must be asked is whether these peaks correspond to any occurrences whatever in the lattice.

In order to answer this question, one may consider the effect produced by the addition of substances possessing an electron affinity, such as biphenyl and naphthalene, to cyclohexane and ethanol. It has been noted that such an addition increases

Table 5.1. Transitions in organic substances

Substance	Phase	Transition temperatures (K)	Radiothermoluminescence peaks (K)	Nature of the transition
Methyl alcohol	cr	157	120, 155	—
Ethyl alcohol	am	110	103–107	—
	cr	110,* 156†	113, 143	*Recrystallization †Melting
n-Butyl alcohol	cr	160*	110, 152	*Phase transition of first order
	am	120–130,*	123	*Glass transition
Benzene	cr	110*	129, 190, 223	*Onset of rotation of the molecule *Phase transition
Hexamethyl-benzene	cr	108,* 135–165†	117, 144, 180, 220	*Phase transition †Onset of rotation of the molecule
Cyclohexane	cr	133–183,* 186,† 218‡	114, 158, 186, 202	*Onset of rotation of the molecule †Crystal latice rearrangement ‡Onset of self-diffusion
1,1-Dicyclo-hexyldodecane	cr	300*	137, 302	*Melting
	am	190*	140, 197	*Glass transition
n-Octadecane	cr	297,* 301†	162, 293	*Phase transition of first order †Melting
Paraffin	cr + am	152, 230–250,* 333†	152, 223, 330	*Glass transition †Melting
Polyethylene	cr + am	173–208,* 236†	123–165, 175–208, 237–248	*Onset of rotation †Onset of segmental motion
Polyisobutylene	am	221*	155, 228	*Glass transition
Teflon	cr + am	168, 295*	150, 295	*Phase transition of first order

Source: Refs. 1–3.

the intensity of radiothermoluminescence and modifies its spectrum but does not change the position of the peaks [2].

The only kind of trap which exists in saturated hydrocarbons is a "hole" formed as the result of electron polarization of the surrounding medium with its induced moments. The energy (or depth) of the trap may be estimated to several tenths of an electronvolt and may be shown to be at maximum if electrons happen to be in a "defect" (cavity) in the lattice. If the substance is polar, the trap may be formed by an abnormal orientation of the dipoles, pointing to a cavity. Such defects may already exist in a rigid medium or may be created by the electron itself if the type of medium permits the reorientation of dipoles. The depth of the trap may then reach 1–2 eV.

Finally, for the medium made up of (or containing in dissolved form) molecules having electron affinity or radicals (these compounds being present initially or, as is the case with radicals, created by irradiation), these entities may intercept the electrons with the formation of ions or of negative ion-radicals. In this event, the depth of the trap is determined by the electron affinity of the compound in question (a range of between 0.4 and 0.7 eV for polyaromatic hydrocarbons and of between 1 and 2 eV for radicals).

Thus it can be concluded that the depth of the traps is different in pure cyclohexane and ethanol, where it is mainly a question of solvated electrons, and in the presence of polyaromatic compounds, where negative ions may be formed. Therefore, electron release from the trap does not consist of electrons evaporating from their traps, but must rather consist of a mechanism able, on the one hand, to liberate solvated electrons (ones trapped in cavities) and, on the other hand, to bring about the rapid diffusion of molecular ions having dimensions on the same order as those of the molecules composing the lattice. We are thus led to believe that all thermoluminescence peaks coincide with any and all reorganizations of the molecular network. The other evidence supporting such a conclusion is provided by the detailed study of thermoluminescence in cyclohexane having undergone varying thermal treatments and the correlation of the temperature of the peaks with crystallographic, calorimetric, and other kinds of data [2]. This study allowed the linking of at least three of the four thermoluminescence peaks from this substance with reorganization of the crystal lattice.

Along with low-molecular-weight substances, the analogy between the positions of thermoluminescence peaks and structural transitions can be readily appreciated in polymers as well. A comprehensive analysis of dilatometric, calorimetric, dynamic-mechanical, and electron-spin resonance data led Boyer to the conclusion that polyethylene exhibits three low-temperature transitions at about 150, 190, and 240 K [4]. As a rule, none of the methods taken separately is sufficient to resolve all three transitions. By means of radiothermoluminescence, however, three transitions whose positions coincide well with Boyer's estimate are observed [5].

It was noted that factors affecting the position of the glass transition and other transitions in a polymer (cross-linking, plasticization, pre-orientation of the polymer film, heating rate, etc.) also affect in a similar way the corresponding peaks on the radiothermoluminescence curve. Moreover, the activation energies for radiothermoluminescence and molecular motion in polymers are in good agreement over the whole range of temperatures studied [6–12].

All this indicates that molecular motion is a determining factor in the destruction of active products of radiolysis in organic substances; the rapid release and recombination of stabilized active particles is possible only by thawing out the molecular mobility in the regions of temperature transitions. The similarity of the activation energies for radiothermoluminescence and molecular mobility over the same temperature range serves as the basis for the hypothesis that the lifetime of charges stabilized on structural elements (or close to them) is in general proportional to the relaxation time of structural elements in question. The specificity of molecular motion, namely, its relaxational nature, governs the kinetics of recombination of charges in amorphous and semicrystalline organic substances as well as the radiothermoluminescence process. The relaxational character of molecular mobility leads to a re-orientation of certain structural elements even at temperatures below those of tem-perature transitions registered by radiothermoluminescence. As a result, an isothermal luminescence occurs after irradiation. An increase in temperature leads to a decrease in the relaxation time of structural elements, to a decrease in the lifetime of charges stabilized on these structural elements, and to an increase in luminescence intensity. Other conditions being equal, the larger the number of structural elements of a given type in a polymer, the more charges will be released on thawing out of the mobility of these structural elements and the higher will be the intensity of the corresponding radiothermoluminescence peak. One may expect that the radiothermoluminescence maxima will be located in the same temperature intervals as transitions observed by the other methods of analysis.

With amorphous polymers, some of the stabilized charges have already recom-bined at low temperatures as a result of the beginning of motion of side groups or small segments of macromolecules in the irradiated sample. The majority of ions recombine, however, near the glass-transition point, thus resulting in the appearance of the most intensive radiothermoluminescence maximum. Figure 5.1 shows the glow curve of a random styrene–butadiene copolymer containing 30 wt% styrene (curve e). Here also the results of evaluation are given of molecular relaxation in this co-polymer by several other methods. The glass-transition temperature of the copolymer was found to be equal to 211 K, as determined from the position of the intensive

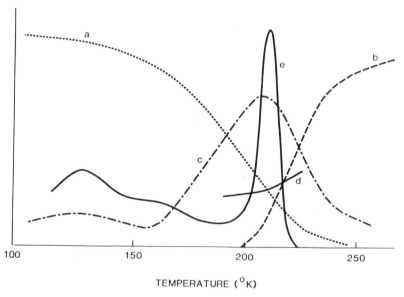

TEMPERATURE ($^{\circ}$K)

Fig. 5.1. Evaluation of the relaxation transitions in a styrene-butadiene random copolymer (30% styrene) by means of various methods: (a) nuclear magnetic resonance absorption line width; (b) tensile compliance; (c) mechanical losses at 0.1 Hz; (d) dilatometry; and (e) radio-thermoluminescence glow curve. After [10], with permission of the Rubber Division of the American Chemical Society.

radiothermoluminescence maximum in the glow curve. This value differs from that measured by other methods by less than 5 degrees.

Similar results were obtained for a great number of amorphous and semicrystalline polymers; the temperature corresponding to the most intensive radiothermoluminescence maximum practically coincides with the glass-transition values determined by relaxational spectroscopy at frequencies of 0.01–0.1 Hz [13]. Besides that, radiothermoluminescence provides one of the most sensitive techniques for evaluating secondary temperature transitions located below the glass-transition and in some cases it also can be used for the analysis of the transitions above the glass-transition.

This rather interesting peculiarity of the radiothermoluminescence process, namely, the occurrence of luminescence flashes at temperatures corresponding to those of temperature transitions, has led, finally, to the creation of a new method of analysis of organic solids and polymers—the radiothermoluminescence method.

5.2 Influence of Impurities on the Glow Curve

The question of the influence of impurities on the radiothermoluminescence process in polymers is of major importance from both theoretical and practical points of view. One of the most significant conclusions agreed on by many researchers is the relative independence of the temperature of the glow peaks and overall shape of the glow curve from the nature and concentration of additives.

Charlesby and Partridge [14] reported that the impurities present in polyethylene have no effect at all on the thermoluminescence emission, since the glow curves of low-density polyethylene specially manufactured without any additives and commercial materials which contain 0.5% antioxidants were virtually identical.

Fleming [15] found that polymethyl methacrylate containing a considerable range of polymerization initiators and chain transfer agents exhibited a glow curve similar to the one for carefully purified methyl methacrylate monomer liquid polymerized by γ-radiation. He also showed that the form of the glow curve is unaltered when polystyrene is doped with 1% by weight of benzophenone, 1,4-cyclohexane dione, or benzophenone oxime (anthracene) [16].

Applying the radiothermoluminescence technique to the study of 1,2-polybutadiene, Bohm [17] found that the luminescence intensity increases with greater purity of the material, whereas the shape of the curve remains unchanged. This suggests that the luminescence species are incorporated into the polymer during the polymerization process or by subsequently occurring grafting reactions in contrast to the apparently excitation-quenching impurities which can be removed by the purification.

A comparison of the thermoluminescence curves for a variety of elastomers and their blends with and without the ingredients of vulcanization led Buben *et al.* [18] to the conclusion that the addition of ingredients has no effect on the shape of the light-emission curves, although it sharply reduces the intensity of luminescence (by the order of 3 to 4).

A comprehensive study of the influence of three different dyes [crystal violet (CV), phenanthrene quinone (Ph), and rhodamine-6G (Rh)] on the glow curves of a polybutadiene (Solprene 233) and a styrene-butadiene block copolymer (Solprene 416) was carried out by Zlatkevich et al. [19]. Doping was performed by two methods: (1) the polymer was kept in the solution (methanol/toluene = 9:1) saturated with the dye for 28 hours and then dried in air, and (2) 0.1 wt% of the dye was roll-milled directly into the polymer.

Similar glow curves have been obtained for the samples doped by swelling and roll milling for all polymer–dopant systems studied. Practically no differences in the overall glow curve shape has been found between the glow curves for undoped samples and those for samples doped with Rh and CV. Ph-doped samples showed a noticeable relative decrease in glow intensity in the temperature region from 120 to 180 K (Fig. 5.2). However, it is important to note that the doping with Ph was accompanied by a change in sample color during and just after preparation. For instance, Solprene 416 roll-milled with Ph turned from yellow (the color of Ph) to pink. Since the polymers with physically dispersed CV and Rh showed glow curves similar to those of undoped samples, the differences in relative thermoluminescence intensity between undoped samples and samples doped with Ph were attributed to the occurrence of a chemical reaction between Ph and other additives contained in or attached to the polymer. It has to be underlined that the position of the main ther-

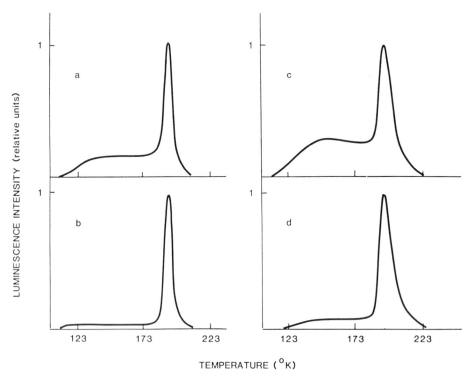

Fig. 5.2. Glow curves for (a) Solprene 233, (b) Solprene 233 doped with Ph, (c) Solprene 416, and (d) Solprene 416 doped with Ph. After [19], © John Wiley & Sons, with permission.

Table 5.2. Temperatures of thermoluminescence maxima for undoped and doped polymers

	Temperature, K	
Sample	Roll milling*	Solvent immersion*
Solprene 233	194	193.5
Solprene 233 and CV	194	193
Solprene 233 and Rh	193.5	194
Solprene 233 and Ph	194.5	194
Solprene 416	195	195
Solprene 416 and CV	195.5	195
Solprene 416 and Rh	195	195.5
Solprene 416 and Ph	195.5	195

*Method of preparation
Source: Ref. 19, © John Wiley & Sons, with permission.

moluminescence maximum (T_g region) was found to be independent of either the polymer–dye combination or the method of incorporation of the dye (Table 5.2).

The picture, however, will not be complete if several contradictory reports are not mentioned here. Using UV light as the excitation source, Linkens and Vanderschueren came to the conclusions that the type of impurity contained in a polymer can play a prevailing role in thermoluminescence and that an observed glow curve is often typical of the dopent–polymer system [20,21]. However, their experiment cannot be considered as a pure radiothermoluminescence test, since the latter requires high-energy ionizing irradiation. Furthermore, differences should be expected between the thermoluminescence curves after exposure to ionizing and UVirradiation, especially for polymer–dopent systems. The first interacts with the whole matrix in a statistically equivalent manner (no selectivity of absorption peculiar to one particular type of molecule), whereas the latter is absorbed only by individual chromophores and aromatic impurities.

Results indicating that the glow curve of polyethylene with an additive does not coincide with the glow curve of the pure polymer were published by Lednev *et al.* [22]. In this study, different additives (phenanthrene, naphthalene, and anthracene) were introduced into polyethylene from the saturated vapor at room temperature, and it was reported that the positions of the individual maxima and the shape of the glow curve are affected not only by the type of the additive, but by variations in its vapor pressure as well.

First of all, it has to be underlined that semicrystalline polymers (especially polyethylene, with its broad variety of different structural formations) are probably not appropriate subjects for studying the influence of additives on radiothermoluminescence glow curve because any changes in the glow curve may result from the structural changes connected with additive introduction. Even when a solvent as an intermediate for the introduction of an additive is omitted and the additive is introduced from the vapor phase, such a procedure cannot be visualized as simply the filling of the free volume of a polymer by additive molecules. Structural rearrangements are expected to take place and can be thought to depend on many factors: the degree of equilibrium of structural formations prior to additive introduction, polymer–additive affinity, the molecular size and concentration of additive molecules, the nucleating efficiency or plasticization ability of the additive, etc. However, there are no obvious reasons for the additive which presumably creates changes in the thermoluminescence curve of a semicrystalline polymer not to do so when introduced into an amorphous material of a similar chemical nature. The additive which serves as a charge trap in a hydrocarbon semicrystalline polymer (only in this case a variation in the thermo-luminescence curve not related to structural changes can be expected) should play the same role in an amorphous hydrocarbon polymer. It has to be noted, however, that the stabilization of one of the charges on additives is the necessary, but not sufficient condition for any change in the glow curve. Meggitt et al. [23] showed that in poly-ethylene doped with biphenyl the most probable radiothermoluminescence mecha-nism involves electron capture by the biphenyl, but luminescence occurs when the cations trapped in the polymer matrix are released and migrate toward the biphenyl anions. The recombination of charges in this case is directly related to "unfreezing" of the molecular mobility in the matrix.

To summarize, one may conclude that at the present time there is no clear evi-dence proving the prevailing role of additives on the shape of the glow curve, pro-vided that introduction of the additive was not accompanied by structural changes in the matrix. On the contrary, the results obtained with a variety of polymers show quite convincingly that additives dispersed into a polymer matrix do not have any significant influence on the position of the thermoluminescence maxima and the over-all glow curve shape. Consequently, radiothermoluminescence can be applied as a method for transition analysis in polymers without interferences owing to possible sample contamination.

5.3   Activation Energy and Methods for Its Estimation

Many methods have been proposed for obtaining activation energies from glow peak temperature and shape. Most of these assume either first- or second-order kinetics, and most also demand a glow peak which is well separated from its neighboring peaks. Furthermore, it is assumed that neither activation energy nor luminescence constant vary during the extent of the peak. Since the methods for evaluating acti-vation energy originally were developed for inorganic substances, it remains to con-sider the meaning to be attached, in the polymer context, to the terms *trap depth* and *frequency factor* found in discussions of thermoluminescence in inorganic materials. For inorganic materials, trap depth is often taken to be the electron affinity of the electron trap, whereas the frequency constant has been visualized as the attempt fre-quency of the electron trying to escape from its trap (the trap is tacitly assumed to be immobile).

Developing the concept that electrons trapped in polymers at low temperature are released not by thermal activation, but by the "unfreezing" of molecular motion,

Partridge [24], instead of considering the electron attempting to gain sufficient energy to escape from an immobile trap, pictured the electron as somewhat loosely bound to a segment of a vibrating (rotating) molecular chain which is attempting to ''shake it off'' when it jumps between equilibrium positions. Such jumping motions are commonly referred to as *primary* and *secondary relaxations* in the context of dielectric and dynamic-mechanical studies.

The relationship between the intensity of thermoluminescence, activation energy, and temperature can be then written as

$$I = \alpha P n v_0 \exp(-E/kT) \tag{5.1}$$

where P is the probability per vibration for a trapped electron to be released, and v_0 is the jump frequency of the molecular chains. Equation (5.1) is identical to Eq. (3.10), except that the frequency factor S in Eq. (3.10) has been replaced by Pv_0 in Eq. (5.1). According to Eq. (5.1), the differences in the glow peak positions are due to the different degrees of molecular motion in different structural regions and the activation energies of the thermoluminescence peaks are in fact the molecular motion activation energies. The fact that Eqs. (5.1) and (3.10) have precisely the same form indicates that the methods of glow curve analysis commonly applied to inorganic materials can be confidently used for organic substances.

5.3.1 Methods Employing Shape Parameters of the Peak

The most widely used methods are summarized in Table 5.3. A number of other methods should be briefly mentioned. Land [25] suggested a method which uses, in addition to T_p, the two inflection points in the thermoluminescence curve rather than the half-intensity temperatures T_1 and T_2. The method based on measurements of T_1, T_p, and T_2 is that of Keating [26]. Maxia *et al.* [27] developed a method for evaluating the activation energy and frequency factor for a multiple-peak glow curve in which the various peaks result from the release of electrons from a single trap and their recombination with various recombination centers. Onnis and Rucci [28] discussed an alternative possibility of obtaining several glow peaks, namely, having several traps and a single recombination center.

It should be noted that all the equations for calculating activation energies assume that the luminescence efficiency of the luminescence centers is independent of temperature. Such an assumption seems reasonable at low temperatures, where radiationless collisional deactivation of the excited states is essentially prevented. However, at relatively high temperatures, phosphorescence quenching may be of significance.

5.3.2 The Initial-Rise Method

Garlick and Gibson [29] suggested a method for activation energy evaluation known as the initial-rise method which is usually considered to be more general than others because it is independent of kinetic order. If one studies Eqs. (3.10) and (3.15), one can see that at the beginning of the glow peak, n changes only slightly with temperature, and therefore, $I \propto \exp(-E/kT)$. Thus plotting $\ln I$ as a function of $1/T$ should

Table 5.3. Various methods of calculating activation energy

Authors	First-order kinetics	Second-order kinetics	Limitations
Grossweiner [42]	$E = 1.51kT_pT_1/(T_p - T_1)$	—	$E/kT_p > 20$
			$3T_1 \exp(E/kT_p)/[2T_p(T_p - T_1)] > 10$
Luschik [43]	$E = kT_p^2/(T_2 - T_p)$	$E = 2kT_p^2(T_2 - T_p)$	$E/kT_p \gg 1$
Kelly and Laubitz [44]	$E = 1.461kT_pT_1/(T_p - T_1)$		$E/kT_p \gg 1$
		$E = 1.763kT_pT_1$	
Halperin and Braner [45]	$E = [q/(T_2 - T_p)]kT_p^2$	$E = [q/(T_2 - T_p)]kT_p^2$	$E > 10kT_p,\ q < 1$
			$1 \leq q \leq 2$

Note: q = the peak symmetry factor
T_p = the peak temperature
T_1 = the temperature on the low-temperature side of the peak at which the luminescence intensity attains its half-maximum value (K)
T_2 = the temperature on the high-temperature side of peak at which the luminescence intensity reduces to its half-maximum value (K)
k = the Boltzmann's constant (eV/K)

Source: Ref [15].

yield a straight line in this region, the slope of which is $-E/k$. The method has further been developed by Gobrecht and Hofmann [30], who used subsequent heating and cooling cycles to obtain the "spectroscopy of the traps."

It should be noted that in the case when phosphorescence quenching is essential, the initial-rise method should yield an underestimated (reduced) activation energy value. As follows from Eq. (3.48), at very low temperatures, quenching will be negligible, but at very high ones, Eq. (3.48) becomes

$$\alpha_T \simeq 1/a \exp(W/kT) \tag{5.2}$$

and this may essentially modify the glow curve and kinetic constants calculated from it. For example, in the extreme case of Eq. (5.2), the form of Eq. (3.10) is now

$$I = (Sn/a) \exp[(W-E)/kT] \tag{5.3}$$

Since the same molecular motion will usually both untrap charges and quench phosphorescence, the activation energy measured by the initial-rise method could vary between E and 0, depending on the temperature of measurement and the phosphorescence content of the thermoluminescence spectrum.

5.3.3   Various Heating Rates

Another group of important methods is that of various heating rates. Upon differentiation of Eq. (3.17) with respect to heating rate, one obtains

$$\frac{dT_p}{d\beta} = \frac{T_p^2}{2T_p + (E/k)} \tag{5.4}$$

From Eq. (5.4) it is clear that an increase in heating rate will always shift the peak maximum toward higher temperatures, and vice versa. Furthermore, this shift will be largest for glow peaks with the lowest activation energy. Bohun [31] and Parfianovitch [32] suggested that a sample should be heated at two different linear heating rates, β_1 and β_2, and that the corresponding peak temperatures T_{p_1} and T_{p_2} should be registered. Equation (3.17) can then be written once for β_1 and T_{p_1} and once for β_2 and T_{p_2}. If one divides these equations one by the other, one gets an explicit equation for the calculation of E:

$$E = [kT_{p_1}T_{p_2}/(T_{p_1}-T_{p_2})] \ln[(\beta_1/\beta_2)(T_{p_2}/T_{p_1})^2] \tag{5.5}$$

Hoogenstraaten [33] suggested the use of several (linear) heating rates; plotting $\ln(T_p^2/\beta)$ versus $1/T_p$ should yield, according to Eq. (3.17), a straight line from whose slope E/k, E is found.

Haering and Adams [34] have shown that for a first-order thermoluminescence peak, the peak intensity is proportional to $\exp(-E/kT_p)$. Thus plotting $\ln I_p$ as a function of $1/T_p$ should give a straight line with a slope of $-E/k$. Another approximate method using various linear heating rates [35] suggests the plotting of $\ln (1/\beta)$ versus $1/T_p$, which should yield a straight line whose slope is E/k. Chen and Winer [36] showed that even for other than first-order cases, plotting $\ln I_p$ or $\ln (\beta/T_p^2)$ versus $1/T_p$ would yield a straight line having a slope of $-E/k$ to a very good approximation.

It is to be noted that even in the case where temperature quenching is essential, utilization of one of the various heating rate methods would yield E rather than $(E-W)$, which is found by the initial-rise method. This is so because the equation corresponding to Eq. (3.17) in this case is

$$\beta(E-W)/kT_p^2 = S \exp (-E/kT_p) \tag{5.6}$$

Thus, by finding E by one method and $(E-W)$ by another, a good estimate of W can be obtained.

5.3.4 Isothermal Decay

The method of isothermal decay enables the measurement of E and S in the first-order case. If one holds a sample at a constant temperature in a range where thermoluminescence appears during heating, one can measure the isothermal decay, which is given by the solution of Eq. (3.10) for the T = constant case as follows:

$$I(t) = nS \exp (-E/kT) \exp [-St \exp (-E/kT)] \tag{5.7}$$

Plotting $\ln [I(t)]$ as a function of t would give a straight line (the occurrence of a straight line ensures the first-order property) the slope of which is

$$M = S \exp (-E/kT) \tag{5.8}$$

Repeating the measurements at various temperatures, one gets various values of M. Plotting $\ln M$ as a function of $1/T$ should give a straight line with slope $-E/k$, thus enabling the evaluation of E and, by substituting into Eq. (5.8), determination of the value of S. This method yields the value of E in the case of exponentially temperature-dependent recombination probability [37].

5.4 Quasicontinuous Distribution of Activation Energies. Temperature Dependence of the Frequency Factor

Analysis of the thermoluminescence process usually assumes one activation energy and one frequency factor for an isolated luminescence maximum, implying that the associated electron traps all have the same activation energy and that the frequency factor is temperature-independent. The assumption of a single activation energy is probably valid for inorganic crystalline dielectrics, but one would expect an activation energy distribution in a largely amorphous organic polymer because of random variations in the surroundings of the electron traps.

Pender and Fleming [38,39] offered a method of numerical analysis of thermoluminescence data when a quasicontinuous distribution of activation energies is present and the luminescence process follows first-order kinetics. Along with obtaining the distribution of activation energies, the method also allows evaluation of the temperature dependence of the frequency factor.

There exist simple experimental tests which will indicate clearly whether a quasicontinuous distribution of activation energies is present in any sample in which luminescence is observed. First, the temperature of the glow maximum intensity is recorded as a function of increasing radiation dose. If it does not decrease steadily, then the luminescence kinetics is first-order rather than a higher order, independent of whether a single activation energy or a distribution exists. Given first-order kinetics, one then carries out a sequence of thermal quenchings. This involves heating the sample until the luminescence intensity reaches a maximum, cooling as rapidly as possible to a temperature well below that of the maximum, and then reheating until the intensity again reaches a maximum. The process is repeated until the intensity level falls to an unworkable level. If successive intensity maxima occur at steadily increasing temperatures, then a distribution of activation energies is indicated; otherwise, a single discrete activation energy value exists. Suppose now that higher-order kinetics are indicated by the increasing-dose experiments. Then it may be shown that provided the probability of retrapping is significant (greater then 0.1) and the proportion of the electron traps occupied at any temperature is very small, then, for a single activation energy value, a plot $\{[I(t)/I_0]^{-1/2} - 1\}$ against t will be linear, where $I(t)$ is the luminescence intensity at time t during an isothermal decay experiment, and I_0 is the initial intensity [38]. This result holds for isothermal decay at any temperature; the form of the decay will be different if a distribution of activation energies exists.

Assuming that in Eq. (5.1) the "luminescence" constant α is temperature-dependent and the constant P is both temperature- and activation energy-dependent, it can be rewritten as

$$I = \alpha(T)P(T, E_j)n \exp(-E_j/kT) \tag{5.9}$$

where v_0 is absorbed into $P(T, E_j)$.

When a quasicontinuous range of activation energies E_j exists,

$$\frac{dn(E_j)}{dt} dE_j = -P(T, E_j)n(E_j) \exp(-E_j/kT) dE_j \tag{5.10}$$

and

$$I(t) = -\alpha(T)\int_E \frac{dn(E_j)}{dt} dE_j \tag{5.11}$$

where $n(E_j)\, dE_j$ is the concentration of electrons in traps with activation energies in the range E_j to $E_j + dE_j$.

Integrating Eq. (5.10) with respect to time, and dropping the subscript j for convenience, one obtains

$$n(E)\, dE = n_0(E)\, \exp\left[-\int_0^t P(T,\ E)\, \exp\,(-E/kT)\, dt\right] dE \qquad (5.12)$$

where $n_0(E)$ is the concentration of electrons in traps in the range E to $E + dE$ at time $t = 0$.

For a glow curve obtained with a heating rate $dT/dt = \beta$, we have

$$I(T) = \alpha(T) \int_{E_{\min}}^{E_{\max}} P(T,\ E) n_0(E)\, \exp\left[-\int_{T_0}^{T} \frac{P(T',\ E)}{\beta}\, \exp\,(-E/kT')\, dT'\right] \exp\,(-E/kT)\, dE \qquad (5.13)$$

where $E_{\min}$ and $E_{\max}$ are, respectively, the lower and upper limits of the activation energy distribution, and T_0 is the initial temperature.

A detailed description of the procedure to be followed in order to obtain the activation energy distribution function $n_0(E)$ and frequency factor distribution $P(T)$ for a well-isolated peak in a glow curve is outlined in [38] and is applied for the evaluation of these parameters for the thermoluminescence maximum in polystyrene. It was shown that the thermoluminescence peak centered at 110 K is characterized by the broad distribution of activation energies (from 0.05 up to 0.45 eV) and frequency factor values (from 10 sec^{-1} at 80 K to 10^6 sec^{-1} at 250 K). The transition at 110 K clearly is a secondary transition (the glass-transition temperature of polystyrene is approximately 370 K). The variations in activation energy and frequency factor in the temperature region of the glass transition may be expected to be even larger, since relaxation times associated with this transition do not follow a Boltzman distribution, but rather the Williams-Landel-Ferry (WLF) equation [40] and activation energy in the glass-transition region is often found to be temperature-dependent. Because of the last factor, it may be beneficial to replace the range of activation energies envisioned in the analysis of secondary transitions by a range of glass-transition values when the glass transition is considered. It has been shown that a range of T_g values exists in several polymers [41].

Taking into account Eq. (3.22), Eq. (5.13) can be written as

$$I(T) = \alpha(T) \int_{T_{g\min}}^{T_{g\max}} P(T,\ T_g) n_0(T_g)\, \exp$$

$$\left[-\int_{T_0}^{T} \frac{P(T',\ T_g)}{\beta}\, \exp\frac{C_1(T'-T_g)}{C_2 + T' - T_g}\, dT'\right] \exp\left[\frac{C_1(T-T_g)}{C_2 + T - T_g}\right] dT_g \qquad (5.14)$$

where $n_0(T_g)\, dT_g$ is the concentration of electrons in traps associated with chain segments which undergo the glass transition in the temperature range T_g to $T_g + dT_g$, and T_0 is the initial temperature.

5.5 The Shape of Thermoluminescence Peaks and Structural Uniformity of a Substance

It has been shown by Luschik [43] that the half-width of the high-temperature side of the elementary thermoluminescence peak δ_2 is connected with the peak-temperature position T_p and activation energy E as follows:

$$\delta_2 = (kT_p^2)/E \tag{5.15}$$

Since first-order peaks are asymmetrical, the half-width of the low-temperature side being about 50% larger than that toward the fall-off of the glow peak [46],

$$\Delta T_0 = 2.5\delta_2 \tag{5.16}$$

where ΔT_0 is the total half-width of the peak. Combining Eqs. (5.15) and (5.16), the half-width of the elementary maximum is

$$\Delta T_0 = 2.5(kT_p^2/E) \tag{5.17}$$

Structural nonuniformity of a substance leads to broadening of the maximum, and it becomes "nonelementary"; i.e., it consists of a group of closely situated maxima which differ slightly in their parameters. If the width of this group ΔT_p appreciable exceeds the width of the elementary maximum, the difference $\Delta T_p - \Delta T_0$ characterizes the nonuniformity of the transition throughout the volume of the specimen.

5.6 The Connection Between Thermoluminescence and Electrical Conductivity in Irradiated Polymers

Differently from static conductivity, which is governed by impurities, conductivity in irradiated polymers is determined by ionization of the matrix and thus has a lot in common with radiothermoluminescence.

Radiation produces the majority of ionizations via secondary and higher-order electrons. Ion pairs formed as a result of ionization are not homogeneously distributed, but are formed in clusters of high-ionization density. An electron is thermolized at a relatively short distance from the positive ion. This thermolization distance in liquids and solids is on the order of 100 Å [47]. At such distances, the Coulomb field around the positive ion is appreciable and the majority of electrons, unless they are trapped, recombine with their parent ion. Such recombination is called *geminate recombination*. There is a probability, however, that the electron released during the ionization will surmount the Coulomb potential well and drift away from its parent ion, avoiding geminate recombination. The G value for electron production is equal to G value for ion-pair production and is usually about 3–4, whereas the G value for nongeminate electrons in condensed phase is 0.05–0.2. Thus something less than 5% of all electrons generated by irradiation are of nongeminate nature [48]. In the absence of an electron scavenging, such nongeminate electrons will eventually undergo recombination with positive ions in distant spurs. If irradiation is carried out at low temperatures and then the substance is heated up, some of the electrons which undergo geminate recombination give rise to luminescence. Nongeminate electrons which are free to move in the polymer may be retrapped, scavenged by chemical impurities, or recombined with other positive ions. However, the distance moved before nongeminate electrons recombine and give rise to luminescence will be far greater than that moved by the geminate electrons, and subsequently, we may expect an appreciable delay between initial detrapping, subsequent retrapping and detrapping, and final recombination for nongeminate electrons.

This must be contrasted with the mechanism of electrical conductivity. Only those electrons which are free to drift under the influence of an applied electric field can be expected to contribute to the observed current. In the absence of any potential difference or thermal gradient across the sample, one would expect not directional, but random charge movement. Under the influence of an applied electric field, the electrical conductivity is enhanced by the directional drift of nongeminate electrons prior to their recombination or retrapping. An electric field also may somewhat ''unbalance'' the geminate recombination, so some of geminate electrons also can contribute to conductivity. The fact that nongeminate electrons can be retrapped and subsequently detrapped again means that the current may flow long after the geminate recombination responsible for thermoluminescence.

The peculiarities of electrical conductivity in irradiated polymers under the influence of a temperature gradient in the sample provide the basis for the so-called thermostimulated-current (TSC) method of the analysis. This method is often used in combination with radiothermoluminescence [48, 49, 54].

5.7   Radiothermoluminescence in Comparison to the Other Methods of Transition Analysis

Radiothermoluminescence possesses numerous advantages over other methods for observing transitions in polymers. Among the advantages are the following:

1.   Speed and relative simplicity of the analysis. With normal heating rates (10°C/min), a run is completed in few tens of minutes.

2.   The thermoluminescence of many irradiated organic substances is so intense that it can be observed with the naked eye. With photomultiplier tubes, it is possible to study the thermoluminescence of speciments whose weight does not exceed a few tenths of a milligram.

3.   Samples of arbitrary configuration can be investigated, e.g., films, pellets, fibers, individual crystals, sliced sections, etc. Because sample shape/size requirements are minimal, strained specimens can easily be investigated. One simple device uses a metal ring to secure the stretched sample on a metal backing plate.

4.   High accuracy and resolution of measurements. The radiothermoluminescence

method gives a continuous curve of luminescence intensity versus temperature, whereas many other methods (mechanical and dielectric spectroscopies, for instance) require repetitive measurements of a certain parameter at different temperatures and then plotting of this parameter versus temperature. Determination of transitions by means of radiothermoluminescence is differential (recording of the peak positions), whereas nuclear magnetic resonance, volume dilatometry, and thermomechanical analysis give an integral curve (the transition corresponds to the inflection point). The temperature of the transition, as given by the peak position T_p, is obtained directly by the radio-thermoluminescence experiment. Because of the narrowness of the peak, T_p usually can be determined with an uncertainty of less than 1°C. This means that subtle effects resulting in shifts of the transition temperature by only 1°C can be studied by the radiothermoluminescence method.

5.   It is well known that multiple transitions are best resolved by a low-frequency test method; transition peaks tend to merge with one another as the measuring frequency is increased. Although in static nonisothermal experiments various molecular motion frequencies usually correspond to transitions located at different temperatures, the range of effective frequencies covered by radiothermoluminescence is somewhere between 10^{-4} and 10^{-1} Hz, thus permitting excellent resolution of overlapping processes.

6.   The half-width of the thermoluminescence peak (ΔT_p) also provides valuable information. If the half-width of the peak appreciably exceeds the half-width of the elementary maximum ΔT_0, the difference $\Delta T_p - \Delta T_0$ characterizes the nonuniformity of the transition temperature throughout the volume of the specimen.

7.   The activation energy of molecular relaxation can be measured over a broad temperature range.

Radiothermoluminescence is especially suitable for evaluating temperature transitions at low temperatures, although there are several reports stating the usefulness of the technique at temperatures essentially above room temperature [50,51].

An inherent disadvantage of the radiothermoluminescence method in its conventional setup is the need to use a high-power source of irradiation. However, instruments with a self-contained irradiation source have been recently introduced [53, 55].

5.8   Instrumentation

The radiothermoluminescence method is noted for the relative simplicity of the installation for registering glow. One of the possible constructions is schematically illustrated in Fig. 5.3. It consists of two parts [52]:

1.   The cryostat
2.   The optical system

The cryostat can be cooled down to $-196°C$ by liquid nitrogen circulation and heated up to 100°C with the heater.

The optical part of the installation consists of a photomultiplier tube and a dia-

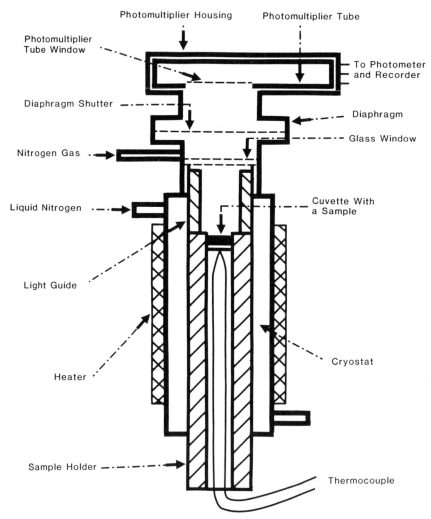

Fig. 5.3. Radiothermoluminescence detector and cryostat. After [52], © John Wiley & Sons, with permission.

phragm with a shutter. The diaphragm shutter is used to establish the "zero light level" of the photometer and to avoid its overloading.

The cryostat is separated from the optical part of the installation by the glass window. To avoid moisture condensation on the window, dry nitrogen is purged over it during the experiment. The spectral composition of thermoluminescence emission can be studied by replacing the glass window with a light filter. After the cryostat is cooled down to $-196°C$, the cuvette with a pre-irradiated sample is placed on the top of a sample holder and is fed into the cryostat, where it is in tight contact with the light guide. The circulation of liquid nitrogen through the cryostat is then shut down, the heater is turned on, and the sample is warmed up at a constant rate. The temperature of the sample is registered by a constantan–copper thermocouple and the intensity of emitted light is recorded continuously.

Irradiation of the sample prior to analysis is usually carried out with γ-rays from a ^{60}Co source, although X-ray and β-radiation sources have been used. The dose rate is not an important factor. All that matters is that the total dose should be from 0.1 to several Mrads. As a rule, irradiation of a substance at a dose of 1 Mrad causes no marked change in its structure. Hence the results can be related to the initial nonirradiated substance.

It has been shown recently that the radiothermoluminescence experiment can be performed using relatively low-energy electrons for irradiation [53]. A beam of electrons having an energy of from 10–50 keV can be incorporated into the radiothermoluminescence installation and does not need any special insulating precautions. An apparatus which combines the irradiation source, cryostat, and optical system is a simple construction. The need to have a vacuum chamber for specimen irradiation and subsequent thermoluminescence analysis is the only increase in complexity required.

The other even more attractive opportunity was provided by the development of a cryogenic scanning electron microscope stage which permits electron irradiation of samples and subsequent thermoluminescence analysis [55]. Irradiation can be performed with either a fixed, defocused electron beam or a scanning beam. The latter permits irradiation of only part of the total specimen and additionally gives more uniform irradiation over the sample. The detection system utilizes a vacuum feedthrough manipulator for positioning the photomultiplier tube before and after sample irradiation. A thin layer of gold (80–250 Å) on the sample surface prior to irradiation is used for elimination of surface charging. It was found that the deposition of up to 250 Å of gold on the sample surface has practically no effect on the overall shape of a radiothermoluminescence glow curve; at the same time, scanning electron microscopy indicated the complete elimination of surface charging. Other features of the system include very short irradiation times (the dose of 1 Mrad can be achieved in seconds) and improved environmental control during heating (N_2, O_2, etc.). Also, varying the electron energy permits control of penetration distance, thus achieving a type of depth profiling not available with other spectroscopic techniques.

Commercialization of the scanning electron microscope attachment containing the cryogenic stage and the vacuum feedthrough manipulator promises to make the radiothermoluminescence method more convenient and widespread.

References

1. Nikolskii, V.G., Buben, N.Ya.: Proc. Acad. Sci. USSR Phys. Chem. *134*, 827 (1960)
2. Magat, M.: J. Chimie. Phys. *63*, 142 (1966)
3. Semenov, N.N.: Pure Appl. Chem. *5*, 353 (1962)
4. Boyer, R.F.: Macromolecules *6*, 288 (1973)
5. Zlatkevich, L.Yu., Crabb, N.T.: J. Polym. Sci. Polym. Phys. Ed. *19*, 1177 (1981)
6. Mindiyarov, Kh.G., Zelenev, L.Yu., Bartenev, G.M.: Polym. Sci. USSR *A14*, 2347 (1972)
7. Nikolskii, V.G.: Pure Appl. Chem. *54*, 493 (1982)

8. Nikolskii, V.G., Zlatkevich, L.Yu., Borisov, V.A., Kaplunov, M.Ya.: J. Polym. Sci. Polym. Phys. Ed. *12*, 1259 (1974)
9. Zlatkevich, L.Yu., Nikolskii, V.G.: Rubber Chem. Technol. *46*, 1210 (1973)
10. Zlatkevich, L.Yu.: Rubber Chem. Technol. *49*, 179 (1976)
11. Nikolskii, V.G., Zlatkevich, L.Yu., Konstantinopolskaya, M.B., Osintseva, L.A., Sokolskii, V.A.: J. Polym. Sci. Polym. Phys. Ed. *12*, 1267 (1974)
12. Nikolskii, V.G., Burkov, G.I.: High Energy Chem. *5*, 373 (1971)
13. Nikolskii, V.G.: Sov. Sci. Rev. *3*, 77 (1972)
14. Charlesby, A., Partridge, R.H.: Proc. R. Soc. *A271*, 170 (1963)
15. Fleming, R.J.: J. Polym Sci. [A2] *6*, 1283 (1968)
16. Pender, L.F., Fleming, R.J.: J. Phys. [C] *10*, 1571 (1977)
17. Bohm, G.G.A.: J. Polym. Sci. Polym. Phys. Ed. *14*, 437 (1976)
18. Buben, N.Ya., Goldanskii, V.I., Zlatkevich, L.Yu., Nikolskii, V.G., Raevskii, V.G.: Polym. Sci. USSR *A9*, 2575 (1967)
19. Zlatkevich, L., Nichols, L.F., Crabb, N.T.: J. Appl. Polym. Sci. *25*, 963 (1980)
20. Linkens, A., Vanderschueren, J.: J. Electrostat. *3*, 149 (1977)
21. Linkens A., Vanderschueren, J.: J. Polym. Sci. [B] *15*, 41 (1977)
22. Lednev, I.K., Aulov, V.A., Bakeev, N.F.: Proc. Acad. Sci. USSR Phys. Chem. *265*, 659 (1982)
23. Meggitt, G.C., Charlesby, A.: Radiat. Phys. Chem. *13*, 45 (1979)
24. Partridge, R.H.: J. Polym. Sci. [A] *3*, 2817 (1965)
25. Land, P.L.: J. Phys. Chem. Solids *30*, 1681 (1969)
26. Keating, P.N.: Proc. Phys. Soc. *78*, 1408 (1961)
27. Maxia, V., Onnis, S., Rucci, A.: J. Lumin. *3*, 378 (1971)
28. Onnis, S., Rucci, A.: J. Lumin. *6*, 404 (1973)
29. Garlick, G.F.J., Gibson, A.F.: Proc. Phys. Soc. *60*, 574 (1948)
30. Gobrecht, H., Hofmann, D.: J. Phys. Chem. Solids *27*, 509 (1966)
31. Bohun, A.: Czech. J. Phys. *4*, 91 (1954)
32. Parfianovitch, I.A.: J. Exp. Theor. Phys. USSR *26*, 696 (1954)
33. Hoogenstraaten, W.: Philips Res. Rep. *13*, 515 (1958)
34. Haering, R.R., Adams, E.N.: Phys. Rev. *117*, 451 (1960)
35. Dussel, G.A., Bube, R.H.: Phys. Rev. *155*, 764 (1967)
36. Chen, R., Winer, S.A.A.: J. Appl. Phys. *41*, 5227 (1970)
37. Wintle, A.G.: J. Mater. Sci. *9*, 2059 (1974)
38. Fleming, R.J., Pender, L.F.: J. Electrostat. *3*, 139 (1977)
39. Pender, L.F., Fleming, R.J.: J. Phys. [C] *10*, 1561 (1977)
40. Williams, M.L., Landel, R.F., Ferry, J.D.: J. Am. Chem. Soc. *77*, 3701 (1955)
41. Bueche, F.: Physical Properties of Polymers. Wiley, New York, 1962.
42. Grossweiner, L.I.: J. Appl. Phys. *24*, 1306 (1953)
43. Luschik, Ch.B.: Dokl. Akad. Nauk USSR *101*, 641 (1955)
44. Kelly, P.J., Laubitz, M.J.: Can. J. Phys. *45*, 311 (1967)
45. Halperin, A., Braner, A.A.: Phys. Rev. *117*, 408 (1960)
46. Chen, R.: J. Mater. Sci. *11*, 1521 (1976)
47. Williams, F.: The Radiation Chemistry of Macromolecules, vol. 1 (ed. Dole, M.). Academic Press, New York, 1972
48. Blake, A.E., Charlesby, A., Randle, K.J.: J. Phys. [D] *7*, 759 (1974)
49. Ranicar, J.H., Fleming, R.J.: J. Polym. Sci. Polym. Phys.Ed. *10*, 1979 (1972)
50. Hashimoto, T., Sakai, T., Iguchi, M.: J. Phys. [D] *12*, 1567 (1979)
51. Blake, A.E., Charlesby, A., Randle, K.J.: J. Polym. Sci. Polym. Lett. Ed. *11*, 165 (1973)

52. Zlatkevich, L.Y., Crabb, N.T.: J. Polym. Sci. Polym. Phys. Ed. *19*, 1177 (1981)
53. Buben, N.Ya., Grishin, V.D., Nikolskii, V.G., Talroze, V.L., Tochin, V.A.: Br. Pat. 1285650 (1972)
54. Mathur, V.K., Brown, M.D.: J. Appl. Phys. *54*, 5485 (1983)
55. Zlatkevich, L., Crist, B.: (*in press*)

Chapter 6

Application of Radiothermoluminescence to the Study of Polymer Systems

6.1 Evaluation of Processes Accompanied by a Shift in Transition Temperatures

6.1.1 Plasticization

Plasticization is widely used in polymer technology when it is needed to increase flow and thermoplasticity of plastic materials. A plasticizer in a polymer having many points of attachment along the polymer chains breaks the attachments and masks the centers of force for polymer–polymer intermolecular attraction by selectively solvating the polymer at these points. This produces much the same results as if fewer points of attachment had been present on the macromolecules originally. It is usually assumed that solvents or plasticizers of different types are attached to the polymer macromolecules by forces of different magnitude and that none are bound permanently.

In the case of a plasticized polymer, additional free volume is introduced with the plasticizer, and the polymer chains are animated by local Brownian motion at temperatures lower than those for a pure polymer. As a result, plasticization is accompanied by a decrease in the glass-transition temperature of the material. The T_g value for the plasticized polymer decreases linearly with increasing plasticizer concentration according to the following relation [1]:

$$T_g = T_{g0} - kw \tag{6.1}$$

where w is the weight percent of the plasticizer, k is a constant for a given plasticizer–polymer system, and T_{g0} is the glass-transition temperature of the polymer. For most polymer–diluent systems, the constant k in Eq. (6.1) can be expressed as follows [2]:

$$k = 2(T_{g0} - T_{g1}) \tag{6.2}$$

where T_{g1} is the glass-transition temperature of the plasticizer.

Although plasticizers decrease the glass-transition temperatures in amorphous polymers, the temperatures of the secondary transitions are affected to a much lesser extent. This indicates that the secondary transitions are determined by local barriers within the molecule (intramolecular barriers) and that they are not sensitive to changes in free volume.

For partially crystalline polymers, plasticization affects primarily the amorphous regions. Nevertheless, the overall effect of the plasticizer on semicrystalline polymers may be of a complex nature and may result in changes not only in amorphous regions, but also in crystalline regions of different degrees of perfection. The crystalline regions can be affected either by direct penetration of the plasticizer into crystal imperfections (if the size of these imperfections permits) or through the interconnection between crystalline and plasticized amorphous regions.

The radiothermoluminescence method was applied to the evaluation of low-temperature transitions in polyethylene kept in liquid hexane for various periods of time [3]. Pronounced changes in the glow curve were noticed, especially in the region of the transition at 244 K. It is interesting to note that depending on how long the polyethylene was held in the liquid hexane, the position of this transition changed nonmonotonically but passed through a minimum; i.e., after a rapid fall, it rose on further holding. The initial shift in the transition to lower temperatures was attributed to the plasticizing action of hexane and was confirmed by the nuclear magnetic resonance data indicating an increase in molecular mobility in polyethylene amorphous regions.

When holding time in hexane exceeded several hours, the transition temperature started to increase, and a prolonged holding led to a shift in this transition up to 263 K. This latter effect was related to removal of the impurities from the polymer. It was noted that the fall in concentration of the impurities increases the distance of diffusion on recombination, which leads to a shift of the maximum to higher temperatures. However, such an explanation is questionable for two reasons. First, it postulates charge stabilization on the impurities. Second, the diffusion mechanism of charge recombination presumes second-order kinetics. Both these assumptions contradict the established mechanism of charge stabilization and recombination in polymers and, in particular, in polyethylene [4]. It thus seems more likely that prolonged holding in hexane promotes secondary crystallization and, as a result, stiffening of tie molecules. The temperature transition around 240 K in polyethylene has been shown to arise from unfreezing of the mobility in interzonal regions, consisting of tie molecules (i.e., chains passing from one lamella to another) [5].

6.1.2 Cross-Linking

Some polymeric systems cannot be used as raw materials because of low dynamic modulus values; i.e., even at low levels of applied stress they undergo large deformations. In such cases, *cross-linking,* i.e., formation of a three-demensional network in a polymer, drastically changes the material properties. Cross-linking is realized in curing of thermoset resins and in vulcanization of rubbers.

In general, since cross-linking minimizes the sliding of chains past one another (viscous flow), amorphous polymers tend to become elastic. As cross-link density increases, the elongation and swelling by solvents decrease, whereas the rigidity rises. Ultimately, an extremely rigid, nonsoluble, nonmelting material is produced. Crystalline polymers, however, may respond to small amounts of cross-linking by a reduction in crystallinity due to increased difficulty in chain orientation. The polymer may become softer and more elastic, and it may melt at lower temperatures. Increased cross-linking beyond this point, however, has the same effects as noted for the originally amorphous polymer. The properties of polymers of intermediate cross-link density are largely determined by the average chain length between cross-links.

One of the physical consequences of cross-linking a linear polymer is a reduction in volume, partly because creation of a network leads to an increase in internal pressure. A major part of the contraction, however, is due to changes in local molecular packing, leading to decreases in occupied and free volume. These changes are reflected in an increase in the glass-transition temperature T_g. This makes possible the use of radiothermoluminescence in the investigation of sulfur- and radiation-induced vulcanization, as well as other processes resulting in the formation of cross-links between molecules.

The vulcanization process represents the formation of a three-demensional network by cross-linking of macromolecular chains. The structure of the network depends on both the type of cross-links and the concentration of chains inside the network, i.e., on the value $1/M_c$, where M_c is the number-average molecular mass of a chain fragment between the neighboring junctions. There are several methods of M_c evaluation. It can be determined by measuring equilibrium swelling or modulus values and from the glass-transition temperature shift according to the following empirical equation [6]:

$$M_c = \frac{3.9 \times 10^4}{T_g - T_{g0}} \tag{6.3}$$

where T_g and T_{g0} are the glass-transition temperatures of cross-linked and un-cross-linked polymer, respectively. It has to be underlined that for many cross-linked systems the shift in T_g is not independent of the chemical composition of the polymer, and this is a complicating factor [7]. As more and more cross-linking agent is incorporated into the network structure, the chemical composition of the polymer gradually changes. The cross-linking agent can be considered as a type of copolymerizing unit. Thus the shift in the glass-transition temperature is made up of two nearly independent effects: (1) the degree of cross-linking, or $1/M_c$, and (2) the copolymer effect. The cross-linking effect always increases T_g and seems to be largely independent of chemical composition, whereas the copolymer effect can either increase or decrease T_g depending on the chemical nature of the cross-linking agent. Equation (6.3) accounts only for the T_g shift due to cross-linking; the shift due to the copolymer effect is not accounted for. However, Equation (6.3) can be applied for evaluation of the degree of cross-linking achieved by radiation where copolymer effect is absent. If copolymer effect has to be taken into account, the theoretical equations derived by DiMarzio [8] and DiBenedetto [9] can be used.

Although in most cases vulcanization increases T_g by no more than 10–15°C, the radiothermoluminescence method can be successfully applied to the study of subtle structural changes produced by cross-linking [10,11]. In particular, it was found that the rate of radiation cross-linking is essentially dependent on the temperature at which the substance is irradiated (Fig. 6.1). The rate of polybutadiene radiation-induced vulcanization increases with the rise in temperature, the most noticeable increase being observed in the glass-transition temperature range. (The T_g value for polybutadiene with 66% vinyl structures is 229 K, as measured by the radiothermoluminescence method [12]). With an increase in the irradiation temperature from 200 to 240 K, the vulcanization is accelerated by a factor of almost 2.5, whereas the cross-linking density yield at temperatures below 200 K and above 240 K is practically temperature-independent.

Besides providing accurate and reliable data on cross-linking density in individual polymers, the radiothermoluminescence technique proved to be most valuable for examination of the vulcanization of homogeneous and heterogeneous polyblends [11,13,14].

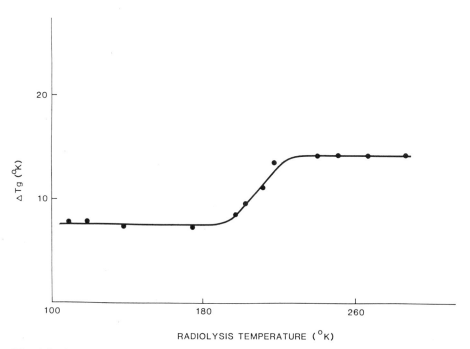

Fig. 6.1. Changes in the glass-transition temperature of polybutadiene rubber (66% vinyl content) vulcanized at a dose of 100 Mrad as a function of radiolysis temperature [10].

6.2   Amorphous Polymers

6.2.1   Glass Transition and Secondary Transitions

It has been known for many years that substances in the glassy state retain some degree of molecular mobility that is detectable by various experimental techniques. These secondary relaxations occur on a time scale many times shorter than that of the main relaxation responsible for the glass transition itself. The presence of such relaxations has generally been associated with the motion of a side group attached to a polymer chain or with a certain type of movement of the chain itself and has been explained in terms of hindered internal molecular modes of motion that remain active even when the molecule as a whole is frozen in place in the glassy matrix [15,16].

The study of glasses made from molecular liquids lacking either a side chain or any other internal degree of freedom capable of giving rise to a relaxation traditionally invoked for polymers and network glasses revealed a remarkable similarity in the relaxation of these glasses to those of polymers [17]. For such glasses, there can be only a short-range, local liquid-like order with no long-range physical connectivity between different parts of the glass, as in the case of polymers. It would appear from these simple considerations that the local structure of the amorphous state other than the distribution of free volume may not play a role in glassy-state relaxation, and the main conclusion was that the molecular mobility seen as secondary relaxations is intrinsic to the nature of the glassy state.

The molecular mobility in glasses can be interpreted in two ways [18]. The first is to suggest that all molecules arranged in a random-network type of a structure in a glass reorient by a small angle, which shows up as the secondary relaxation, and by a large angle, which gives rise to the main relaxation (the small-angle reorientation being a necessary prior step for the large-angle jump). The second interpretation is to assume the presence of two types of environments, one in which the structure is relatively loose and the molecules can undergo hindered rotation and another in which the structure is closely packed and a cooperative rearrangement occurs. Glasses are thought to be composed of aggregates of high density and low energy which lack the symmetry needed for long-range periodic packing. These aggregates are termed *amorphous clusters*. The structure of a glass consisting of such clusters would require relatively loose packing of atoms or molecules in the interstices between the amorphous clusters. In this interpretation, the main relaxation arises from the rearrangement of molecules in the clusters, and secondary relaxation results from that of the molecules in the interstices between the clusters. In contrast to the random-network model, in which all the atomic or molecular units are in an essentially equivalent environment, the amorphous-cluster model implies the presence of two different environments in the structure.

Interestingly, the loss peak of the local main-chain motion and that of the glass transition are clearly separate. There is no gradual transition from a local-mode motion to the glass transition by a gradual enlargement of the moving group with increasing temperature. It follows that there is a distinct difference between the two types of motion responsible for the secondary loss peak and for the glass transition respectively.

Although it does not seem possible to resolve the relative merits of the two structural models by an experimental method, the almost universal presence of secondary relaxation in glasses suggests that two distinct types of molecular environments are also universal and therefore lends support to the amorphous-cluster model.

Evaluation of a variety of amorphous polymers by the radiothermoluminescence method revealed the presence of both the glass and secondary relaxations [19–22]. The former are usually represented by a relatively intensive and narrow peak, whereas the latter are usually represented by a much broader and less intensive maximum.

6.2.2 Changes in Radiothermoluminescence Associated with Induced Crystallization

Some polymers which are regarded as amorphous materials crystallize as a result of stretching. Stretching and the orientation which accompanies it involve molecular arrangements. The higher the orientation, the more mutually parallel are the molecules and the smaller is the average angle formed by them with the axis of orientation. The process of axial orientation may result in crystallization both by orienting the molecules and by bringing them closer together, enabling crystallites to form from formerly amorphous regions. The greater the forces arising between the interacting groups in neighboring molecules, the more groups available, the more regular their disposition along the chain, and the closer to one another these macromolecules can get, the stronger are the intermolecular bonds, resulting in more facile polymer crystallization. Prolonged holding of the material at certain temperatures also can induce crystallization. If there is sufficient motion, random crystallization of nearest neighbors may occur without any redistribution of chain segments. This process was considered by Wunderlich [23] to be "cold crystallization." Crystallites produced by stretching usually occur with their chain direction preferentially oriented parallel to the axis of elongation. This is in contrast to the crystalline texture that results when the transformation is induced merely by cooling. In the latter case, the crystallites are, on average, randomly arranged relative to one another. The crystallization effect can be observed most readily in elastomers with stereoregular macromolecules (polybutadiene, polyisoprene, polyisobutylene, natural rubber). The degree of crystallinity for polybutadiene, for example, can reach 45% [24].

cis-Polybutadiene and polyisobutylene in the amorphous state are characterized, as is usual for amorphous polymers, by secondary relaxation(s) and by the main relaxation in T_g region (Fig. 6.2). Holding of these polymers at low temperatures (263 K for polyisobutylene and 218 K for polybutadiene) induces crystallization and brings about noticeable changes in the glow curves: a decrease in the intensity of the maximum in the glass-transition region and the appearance of a new peak located on the temperature scale between the secondary and glass relaxations [24].

The appearance of the new thermoluminescence maximum at relatively high temperatures essentially exceeding the temperature range of secondary amorphous relaxation indicates the involvement of relatively large molecular sequences and reveals a rather defective crystal structure.

The results shown in Fig. 6.2 were obtained for polymers crystallized at temperatures providing a maximum crystallization rate. Either an increase or a decrease in

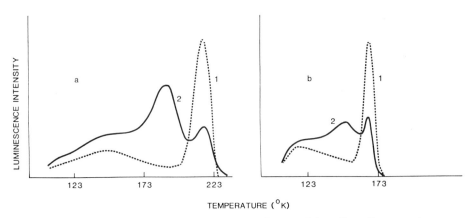

Fig. 6.2. Glow curves of *(a)* polyisobutylene and *(b)* *cis*-polybutadiene (1 = amorphous samples; 2 = semi-crystalline samples [24].

crystallization temperature above or below the optimum value results in a decrease in the crystallization rate. Polymers crystallized for the same period of time at various temperatures are expected to develop a maximum degree of crystallinity when crystallized at the maximum crystallization-rate temperature, and the degree of crystallinity is lower for crystallization temperatures both higher and lower than the optimum value.

It was observed that polyisobutylene slowly crystallizes even at room temperature, although samples had to be stored at ambient conditions for several months for the effect to be noted. The influence of the crystallization temperature on the relative intensities of the T_g peak and the peak appearing as a result of *cis*-polybutadiene crystallization (T_c peak) are shown in Fig. 6.3. As expected, the I_{Tg}/I_{Tc} ratio has a minimum value at 218 K, the maximum crystallization-rate temperature. Assuming a direct relationship between the degree of crystallinity and the intensity of the T_c peak, the former can be thought to be proportional to the $I_{Tc}/(I_{Tc} + I_{Tg})$ ratio.

The effects of crystallization on the glow curve described above are even more pronounced when crystallization is achieved by stretching [26]. The glow curve of stretched polyisobutylene shows a sharp new peak which is also characteristic for samples crystallized by cooling. This peak is obviously of the same nature in both cases and is due to crystallization. One can thus conclude that formation of the additional peak on the glow curve of the apparently amorphous polymer is evidence of its crystallization. The T_c peak of the stretched polyisobutylene is larger than that of the sample crystallized by cooling, and it increases with increases in the degree of deformation. Deformation that causes considerable crystallization of the polymer shifts T_g toward higher temperatures (Table 6.1). T_g elevation due to an increase in crystallinity is linked with a change in macromolecular chain orientation in the amorphous zones when deformation of intercrystalline bonds reduces the number of feasible macromolecular conformations. If one destroys the crystalline zones produced by stretching or cooling by heating the samples to 100°C and holding them at this temperature for 5–10 minutes, the T_c peak will not be present on the subsequently

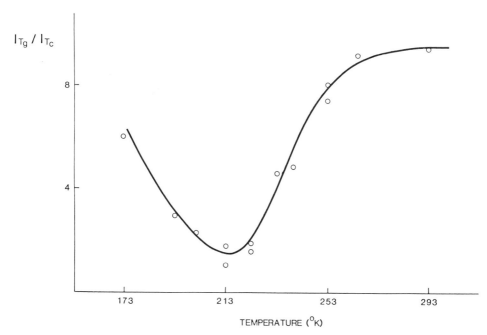

Fig. 6.3. Relative intensities of T_g and T_c transitions in *cis*-polybutadiene as a function of crystallization temperature (crystallization time = 1 hour) [24].

obtained glow curve and its shape will become the same as that of the original amorphous polyisobutylene.

The radiothermoluminescence findings were confirmed by nuclear magnetic resonance and X-ray diffraction studies. X-ray analysis of polyisobutylene showed that

Table 6.1. Transitions in polyisobutylenes with induced crystallinity

Storage temperature (K)	Storage time (days)	Chilled samples			Stretched samples	
		T_g (K)	T_c (K)	Deformation (%)	T_g (K)	T_c (K)
273	2	221	—	0	220.5	—
	7	219	—	100	220	193
	11	218	189	200	219	190
				400	219	190
268	2	221	—	600	221	186
	7	220	197	800	223	187
	11	218	195	1000	229	200
263	2	221	193	—	—	—
	4	219	192	—	—	—
248	2	221	—	—	—	—
	7	222.5	—	—	—	—

Source: Ref. 26, Pergamon Journals, Inc., with permission.

under stretching, its macromolecules form a crystal structure in the orthorhombic system with unit-cell dimensions $a = 6.94$, $b = 11.96$, and $c = 18.63$Å. The unstretched polyisobutylene sample was characterized by a broad amorphous halo, whereas the deformed sample had a narrow reflection due to the crystalline structure. The reflection width was found to be independent of the orientation of the stretching axis relative to that of the X-ray beam. Two minima reflecting the main and secondary relaxations were observed on the nuclear magnetic resonance spin-lattice relaxation time–temperature curve for the unstretched sample, whereas the stretched sample also exhibited a poorly resolved intermediate minimum. The samples crystallized by cooling, however, did not show any noticeable changes, most probably because of insufficient sensitivity of the nuclear magnetic resonance technique.

For polymers, a characteristic peculiarity of the orientation imposed by stretching is that it promotes crystallization and that crystallization in a previously oriented system diminishes the stress. For any stress likely to be borne by amorphous chains, the length of the randomly coiled molecule projected on the axis of orientation is considerably less than its length in the crystalline state. Hence, for oriented systems, melting results in a contraction and crystallization results in an elongation. This behavior, which reflects one of the unique properties of polymer chains, results from their configurational versatility. Macroscopic dimensional changes and changes in the stress exerted can therefore be coupled with and related to the crystal–liquid phase transition. In a well-oriented crystalline polymer, a significant amount of superheating may be required to initiate melting and observe shrinkage.

In order to analyze properly the melting of an oriented system, it must be ascertained whether the process is reversible; i.e., whether oriented crystallites are formed on subsequent recrystallization. This is of concern because it is possible that the original oriented crystalline state will not be regenerated. The possible nonequilibrium aspect of the melting of an oriented polymer results in a significantly higher melting point than that assigned to the equilibrium melting temperature, and a rise in the nonequilibrium melting point with the increased degree of orientation should be expected.

Evaluation of the type (uniaxial or biaxial) and extent of deformation on temperature transitions in polychloroprene rubber by the radiothermoluminescence method gave results essentially similar to those described earlier for polyisobutylene and polybutadiene [27]. Two closely situated maxima were observed in the temperature region of the glass relaxation (from 220 to 260 K) whose positions and intensities were dependent on the degree of orientation (i.e., the degree of crystallinity developed during orientation). The intensive low-temperature maximum at 150 K (secondary relaxation) was found to be independent of the extent of deformation.

In addition to the relaxation maxima in the secondary- and glass-transition regions, the glow curves for crystallized polychloroprene exhibited a maximum in the temperature region of 300–340 K which was attributed to the melting of crystallites. The position of this maximum was found to depend strongly on the type and extent of deformation. The observed changes indicated that the melting temperature increased with increases in the degree of deformation and that for the same degree of deformation, biaxially oriented samples melted at higher temperatures than uniaxially oriented ones. Such a behavior is in accordance with the basics of nonequilibrium melting of polymers crystallized by orientation discussed earlier.

6.2.3 Individual Polymers

6.2.3.1 Polybutadienes

It is well known that the chain microstructure of a polymer influences its relaxation behavior. A case in point is the diene polymers, which can contain essentially different concentrations of cis, trans, and vinyl structures.

The glow curves of three polybutadienes having various vinyl contents are shown in Fig. 6.4. For all the materials, there are two temperature ranges in which the luminescence intensity increases rapidly. The first is 130–160 K. In this temperature range, the glow curves have complicated shapes. For samples A and B, two adjacent maxima are superimposed, whereas sample C exhibits a single, poorly resolved maximum. The second temperature range in which there is a sharp increase in the intensity of luminescence is 160–273 K. In this range, there is a clearly defined single maximum whose position shifts to higher temperatures when the vinyl concentration increases [19].

The presence of the maxima in light emission during warming to room temperature is an indication that an accelerated liberation of electrons from traps takes place at the temperatures at which these peaks appear. Increases in the rate of detrapping can occur because the thermal energy kT is sufficient to lift an electron over the barrier potential of the trap or because the trap itself is eroded by the onset of structural changes occurring at this temperature. As was shown earlier, the latter process predominates in polymer systems.

At least three different types of electron traps can be proposed for polybutadiene [21]. Consistent with the free-volume concept, it can be assumed that fluctuations in density exist in the frozen matrix, manifesting themselves as cavities. Free electrons trapped in some of these abundant low-lying energy wells should easily escape as a result of the supply of only modest amounts of energy. The removal of electrons from deeper cavity traps would require more energy; however, this could be accomplished even at low temperatures by the onset of vinyl group rotation about the C–C bond which links this group to the main chain or by a rotation of short-chain segments in "crankshaft" configurations. The latter should be possible well below T_g and should have a low activation energy. Also contributing can be the vibration of chain segments or submolecules about their equilibrium positions.

A second possibility is for the electron to be attached to double bonds having a positive electron affinity. Since unsaturation and, in particular, vinyl groups are present in polybutadienes, they too represent available electron traps.

Finally, transient intermediates as radicals and radical-ions formed during the primary irradiation must be considered as traps. Since the ionization potential of π electrons is considerably lower than that of σ electrons in the aliphatic chain, it can be assumed that the radical-ions formed by irradiation are predominantly located in the vinyl groups [21].

An unequivocal correlation of the peaks shown in Fig. 6.4 with the types of traps present in polybutadiene is not possible at this time. However, the low-temperature transitions in the 130–160 K region (secondary relaxations) most probable should be assigned to cavity traps which are destroyed by motions involving pendent vinyl groups and short segments of the backbone structure. As Fig. 6.4 illustrates, the

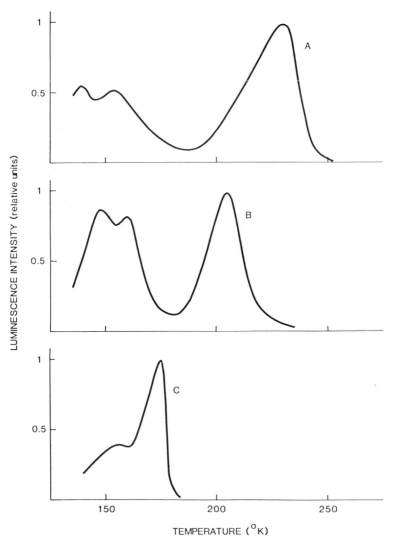

Fig. 6.4. Glow curves of polybutadienes with different vinyl concentrations: *(A)* 66%; *(B)* 40%; and *(C)* 5%. After [19], Pergamon Journals, Inc., with permission.

intensity of secondary relaxation is larger for polymers with high vinyl concentrations. This is predictable, since a high vinyl content inevitable results in a structure with a larger concentration of defects.

Besides secondary relaxation around 160 K, a fairly broad peak at 118 K has been reported for high-vinyl-content polybutadiene [21]. In this study, irradiation was carried out at 90 K, and thus the difference between the irradiation temperature and the position of the first low-temperature maximum was only 28 K. It must be borne in mind that at the beginning of the heating period the intensity of the radiothermoluminescence increases independently of the presence of a transition and that the initial rise in emission is due apparently to the increase in the rate of interaction of the ionized states on heating in comparison with their rate of interaction at the temperature of irradiation. Thus in some cases the first low-temperature maximum (around 90–120 K) is only apparent; its position and shape are determined by the temperature and regime of irradiation. One of ways to discriminate between a true relaxation peak and an apparent peak not directly related to relaxation is to perform irradiation and the start of subsequent heating at still lower temperatures (e.g., at liquid helium temperature instead of liquid nitrogen temperature).

Along with secondary relaxation peaks, each of polybutadienes exhibits the most intensive maximum at temperatures approaching the glass transition. It is believed [21] that all the ion recombinations which could take place through detrapping of the electron and subsequent migration toward the positive charge have occurred. Remaining in the sample are trapped electrons which cannot be liberated by supplying energy of the amount of kT or by the erosion of traps induced by the onset of local motion. While it is possible that certain cavity traps remain intact up to T_g, it is more likely that these deep traps are associated with free radicals, which are known to possess a high electron affinity. Charge recombination involving these groups then occurs at T_g, not by detachment of the electrons, but by a physical approach of the two molecular ions.

Polybutadiene samples with greater vinyl concentrations exhibit higher T_g values. Moreover, there is a direct proportionality between T_g values and vinyl concentration (Fig. 6.5).

The T_g position for all polybutadienes depends on the heating rate, shifting to higher temperatures when heating rate is increased. By assuming a Boltzman relation for the ion-pair recombination-rate constant [Eq. (3.17)], the largest temperature shift is expected for the transition (peak) with the lowest activation energy. Partridge [4] applied Eq. (3.17) to the results obtained by Alfimov and Nikolskii [19] on the variations of glass-transition glow peak temperature with warming rate. Activation energies of about 2 and 1.5 eV were obtained for samples with 66 and 5% vinyl concentrations respectively. These values are in agreement with Alfimov and Nikolskii's own evaluation in accordance with the relation

$$1/T_p = C_1 - k/E \ln \beta \qquad (6.4)$$

where C_1 is a constant, and β is the heating rate. Relation (6.4) follows from the Eq. (3.17) when $E/2kT_p \gg 1$.

Although both Eqs. (6.4) and (3.17) give similar activation energy values, utilization of the Eq. (3.17) is preferable because along with the activation energy, it allows evaluation of the frequency factor. According to Partridge's estimate, the frequency-factor values are about 10^{45} and 10^{42} sec^{-1} for high- and low-vinyl-content polybutadienes, respectively. These values are very high and exceed by far a limiting frequency of 10^{12} sec^{-1}, which corresponds to a transition with zero activation entropy [28]. It should be remembered, however, that we are dealing with the glass transition, i.e., with a transition of a complex nature in which a large activation entropy change is expected. Activation entropies of 60–100 eu have been reported for some relaxations in polymers and even in low-molecular-weight solids [29]. Thus

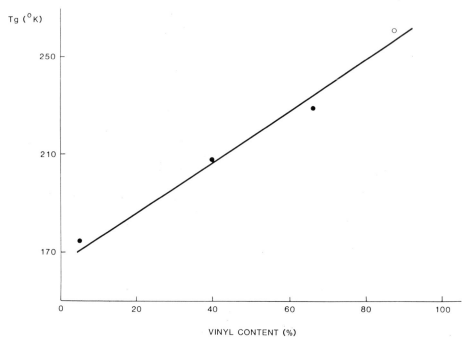

Fig. 6.5. T_g dependence on vinyl concentration for polybutadienes. Solid circles, ref. 19; open circle, ref. 21.

high frequency-factor values on the order of 10^{40} sec^{-1} are not unrealistic and first of all indicate a high degree of complexity in the motion associated with the relaxation.

Glass transitions have large activation energies and activation entropies and are clearly highly complex relaxations. Nevertheless, there is evidence that they may consist of a distribution of simple component relaxations [15]. When a glass transition is considered, it is convenient to replace the range of activation energies by the range of glass-transition values. Assuming that $C_2 \gg T_p - T_g$ in the Williams-Landel-Ferry (WLF) type equation (3.23), it can be expressed as

$$(C_1/C_2)\,(T_p - T_g) \simeq \ln\,(C_1/SC_2) + \ln\,\beta \qquad (6.5)$$

The evaluation of C_2/C_1 ratio by plotting T_p against $\ln\,\beta$ gave 2.2 and 1.7 for low- and high-vinyl polybutadienes, respectively [4]. The constants C_1 and C_2 of the WLF equation are "universal constants" with values of about 17 and 51, respectively, so C_2/C_1 should be about 3; the extent of agreement with the experimental results is quite good, especially because the definition of T_g itself is rather vague [2]. The transition takes place not at a constant temperature, but over a temperature range of several degrees, and none of the abrupt changes in thermodynamic parameters associated with a first-order phase transition is observed in the glass-to-rubber transition. A technique frequently adopted to "define" T_g is to extrapolate two plots of the specific volume of the polymer, one above and one below the transition, and to take T_g as their intersection [176]. Even then, difficulties arise from slight shifts of the intersection temperature, depending on the heating or cooling rate employed [30].

It is not really possible to obtain the frequency factor S because the precise value of T_g is not known (its practical definition is to some extent a matter of experimental approach), and thus, the value $T_p - T_g$ is subject to considerable error. But assuming $T_p - T_g$ be a few degrees only, the value of S will be quite modest, and essentially less than the value obtaind by use of Eq. (3.17).

It is always of interest to compare relaxation parameters obtained by different methods; since this might provide some additional information. In the study conducted by Knappe *et al.* [31] on *cis*-polybutadiene (98% 1,4-*cis*), the activation energy of the glass relaxation was found to be 1.35 eV by the method of initial rises and only 0.43 eV by the method of various heating rates. Such a discrepancy cannot be due to quenching (phosphorescence quenching, if taking place, would give just the opposite effect). As the matter of fact, close activation energy values obtained by Nikolskii when the method of initial rises [32] and the various heating rates method [19] were applied indicate that phosphorescence quenching is of no importance in this particular case. Since the activation energy in the vicinity of 1.3 eV has also been confirmed by other reserchers [33], it is believed that the value of 0.43 eV is underestimated. The most probable reason for this seems to be the insufficient accuracy of the measurements. As was shown by Alfimov and Nikolskii [19], the T_g peak position on the glow curve shifts to higher temperatures with increases in warming rate, as expected, but only by about 6 degrees for a warming-rate increase of 30 times. In the study performed by Knappe *et al.* [31], the heating rate varied from 1 to 3 degrees/minute, i.e., by 3 times, which is not adequate to ensure a reliable activation-energy evaluation.

6.2.3.2 Polymethyl Methacrylate

The most comprehensive study of this polymer was performed by Fleming [34]. The radiothermoluminescence glow curves could be divided into two main categories: those exhibiting a single, isolated peak at about 162 K, and those showing in addition to this peak a shoulder with maximum intensity at 239 K. In connection with the peak at 162 K, the following luminescence mechanism was proposed: It was postulated that electrons displaced by the incident ionizing radiation at liquid nitrogen temperature are loosely bound to the main-chain methyl groups of the polymethyl methacrylate molecules. As the sample warms up, this group will commence rotational motion, and consequently, the trapped electrons will periodically occupy spatial positions in which considerable overlap occurs in their methyl group-based orbitals and in some unoccupied orbitals in appropriate luminescence centers. Radiative combination will then occur, provided the energies of the trapped electrons are compatible with their energies in the orbitals based on luminescence centers.

The shoulder at 239 K was assigned to recombination of electrons at structural defects in a few small crystalline regions. Although the appearance of the poorly resolved maximum around 239 K was unpredictable and inconsistent (out of a total of some 75 glow curves, only 8 gave an unambiguous indication of the existence of this peak), it is interesting to note that this transition, as observed by radiothermoluminescence, has been reported by the other researchers [35] as well.

Fleming [34] measured the activation energies of the glow peaks by a number of different methods and obtained the value of 0.084 eV for the peak at 162 K. This peak certainly followed first-order kinetics, and the frequency constant evaluated from the Eq. (3.17) was about 1.6 sec^{-1}. The low activation-energy and frequency-constant values clearly indicate the "simple" nature of this transition, whose temperature and activation energy are close to the values for rotation of the main-chain methyl groups as measured by nuclear magnetic resonance and dynamic-mechanical experiments. It was not possible to determine the kinetics of the 239 K peak with certainty, but an activation energy of 0.43 eV followed from a first-order assumption.

6.2.3.3 Polystyrene

The polystyrene glow curve has just one glow peak with a maximum at about 120 K [36,37]. The thermoluminescence emission in this temperature region was attributed to the onset of thermally activated oscillation of phenyl groups. This motion is assumed to effectively destroy the electron traps formed by two or more neighboring pendent phenyl groups immobilized at the irradiation temperature (77 K).

Quasicontinuous distribution of electron-trap activation energies ranging from 0.05 to 0.45 eV and the frequency factor ranging $10-10^6$ sec^{-1} were demonstrated [38]. These values compare adequately with the activation energy of 0.21 eV and the frequency factor of 8×10^5 sec^{-1} evaluated by Partridge [39] using the initial-rise and various heating rate measurements, respectively.

It was noted that the longer the delay between the end of irradiation and the commencement of heating, the higher is the temperature at which the initial rise of the thermoluminescence appeared; when the delay was increased from 10 to 30 minutes, the luminescence peak appeared 5 K higher [38]. This observation suggests that the activation-energy distribution of the available traps extends even to lower values, but that traps with lower activation energies than 0.05 eV are empty at 80 K. It would be necessary to irradiate the samples at much lower temperatures in order to investigate this possibility.

The thermoluminescence decay did not follow any simple time dependence, for example $I \propto t^{-1}$ or $I \propto \exp(-t)$, neither at 90 K [37] nor at 120 and 150 K [38]. However, when the quasicontinuous distributions of activation energies and the frequency factors were put into consideration, a good agreement was obtained between the calculated and experimental isothermal decay plots.

6.3   Semicrystalline Polymers

6.3.1   Radiothermoluminescence and Crystallinity

Since the mechanism of radiothermoluminescence in organic substances is directly related to molecular mobility, one would expect to observe noticeable differences in the character of thermoluminescence between amorphous and crystalline regions in the same material. Indeed, the evaluation of substances which were frozen in the form of both a crystal and a glass as well as in the two-phase, amorphous-crystalline state showed that the glow curve reflects even a slight change in the degree of crystallinity [40]. Substances such as toluene, ethylbenzene, tetralin, vinyl acetate, and 1,1-dicyclohexyldodecane vitrify on rapid cooling and crystallize on slow cooling to 77 K. All crystalline samples exhibited an intensive luminescence which continued until they melted. The luminescence of the amorphous samples ceased appreciably earlier. An example of this kind is shown in Fig. 6.6. On warming an amorphous sample of 1,1-dicyclohexyldodecane, luminescence flashes are observed at 140 and 197 K, representing secondary and glass transitions, respectively. The glow curve of crystalline 1,1-dicyclohexyldodecane, however, has no special features in the vicinity of 197 K. After exhibiting secondary transition around 140 K, it continues more or less smoothly up to higher temperatures, and the next flash of luminescence is observed only at the point where the crystal melts (302 K), i.e., at a temperature more than 100 degrees higher than the temperature of the last flash of luminescence for the amorphous sample. Figure 6.6 also shows the glow curve for a finely crystalline sample of 1,1-dicyclohexyldodecane, a white opaque substance formed by the rapid crystallization of the supercooled liquid. In this case, the maximum at 197 K is also absent. The secondary relaxation ($\sim$140 K) is enhanced, but the luminescence intensity in the fusion region is weak, and instead of a single maximum, there are two closely situated, poorly resolved maxima. In general outline, this form of curve corresponds to uniform but highly defective crystals. Holding the finely crystalline sample at 300 K gradually leads to an increase in the size of the crystallites. The shape of the glow curve is altered correspondingly, and prolonged treatment leads to a glow curve which coincides with the curve b in Fig. 6.6.

Rapid cooling of a mixture of 1,1-dicyclohexyldodecane liquid containing fine crystallites gives a heterogeneous two-phase system (Fig. 6.6d). Crystalline and amorphous regions in such a system behave essentially independently—the intensities of the thermoluminescence maxima at 302 and 197 K are proportional to the crystalline and amorphous contents, respectively. Evaluation of the degree of crystallinity from the radiothermoluminescence data is pretty straightforward for low-molecular-weight organic substances which can be obtained in completely amorphous and completely crystalline states. Nevertheless, the perfection of the crystalline phase may vary from sample to sample, and even for a simple two-phase system, the degree-of-crystallinity concept does not define the system thoroughly. The situation is even more complicated with semicrystalline polymers which cannot be approximated by a two-phase amorphous-crystalline model, in which case none of the transitions can be fully attributed to either the amorphous or the crystalline phase. At best, very qualitative estimates can be made. Thus, for polyethylenes, there were some indications that three thermoluminescence peaks appear to be related to crystallinity, the lowest temperature peak ($\sim$135 K) being most prominent with polyethylenes of high density (high crystallinity) and two other peaks ($\sim$185 and 230 K) being more pronounced in polyethylenes of low density (branched) [36]. From the comparison of radiothermoluminescence data for polyethylene single crystals and single crystals fused in a vacuum, it was concluded that the crystallinity rather than other properties does in fact determine the relative importance of the thermoluminescence peaks [41].

Not only the degree of crystallinity, but also some peculiarities of the internal

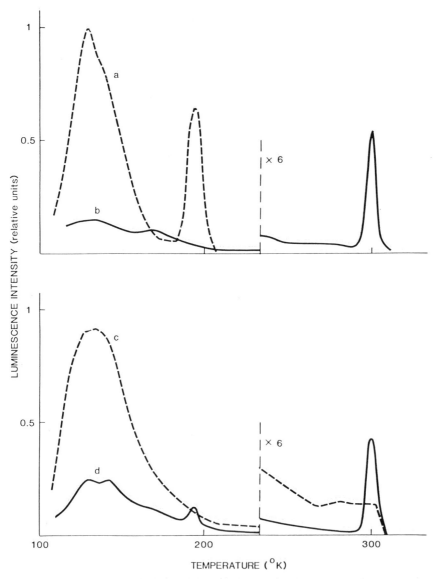

Fig. 6.6. Glow curves of 1, 1-dicyclohexyldodecane: *(a)* glass; *(b)* crystal; *(c)* finely crystalline sample formed by rapid crystallization of the supercooled liquid; *(d)* two-phase amorphous-crystalline sample (15–20% amorphous phase) [40].

crystalline structure are reflected by radiothermoluminescence. None of the odd alkanes, for example, exhibits an oxygen peak in the glow curve, whereas each of the even ones does [42]. This effect is thought to originate in the different crystalline structures of the even and odd paraffins; the even paraffins have triclinic crystal symmetry, whereas the odd paraffins are orthorhombic. Crystal structure differences probably affect both the type of charge traps in the material and the ease with which molecular motion can accomplish untrapping from these traps, as well as, perhaps, the mobility of free charges within the matrix.

The radiothermoluminescence study of polyoxymethylene samples with various morphologies showed significant differences between the glow curves in both shape and intensity [43]. The needle-like crystals exhibited a sharp peak at around 456 K whose half-width was as narrow as 0.5 K, and the temperature position agreed with the material's melting point. For fibrillar crystals, a sharp luminescence peak similar to that from needle-like crystals appeared at around 451 K. The light intensity was small, however, about half that from needle-like crystals. The temperature again corresponded to the melting temperature. Compared with those extended-chain-type crystals, folded-chain crystals showed only very weak luminescence around their melting temperature, and a broad emission maximum was predominant in a wide temperature region between 333 and 443 K. The latter temperature interval corresponds to so-called crystalline dispersion [44] or the occurrence of molecular rearrangement. In the case of extended-chain (needle-like or fibrillar) crystals, the crystalline lattice is considered to be so rigid that no significant molecular motion occurs until the melting temperature is reached. However, molecules in folded-chain crystals may become mobile, in terms of untrapping of electrons, within the crystalline dispersion region.

It was concluded by the authors that radiothermoluminescence provides information about the crystalline lattice which is not obtainable by X-ray or other methods [43]. Later, a similar conclusion was reached when extended-chain and folded-chain polyethylene crystals were studied [45].

6.3.2 Relaxation Transitions in Semicrystalline Polymers

In semicrystalline polymers, relaxation transitions are usually assigned to the amorphous and crystalline phases by the effect of change in crystallinity on the particular transition. Most interpretations have been in terms of an amorphous-crystalline two-phase model of polymer morphology using the degree of crystallinity (as determined by density or some other means) as the primary parameter for comparison of different samples. The effort has gone primarily into inferring the phase in which the relaxation process occurs and the probable type of relaxing species. In many cases, however, such an assignment is not straightforward, since polymer samples can be at the same level of crystallinity but have different morphologies, and vice versa.

Use of the words *crystalline* and *amorphous* implies division of the polymer sample into regions which are wholly one or the other; the possible existence of regions of intermediate order is ignored. Clearly, the concept of degree of crystallinity assumes the existence of a two-phase system. The properties of each phase are assumed to be independent of the presence and amount of the other. The crystalline component is assumed to possess an ideal crystal lattice without defects, and the amorphous one is just supercooled melt with random chain orientation.

Obviously, such a model may be too restrictive for highly crystalline polymers. The polymer sample cannot be either completely crystalline or completely amorphous. By its very nature, a polymer liquid cannot be considered to be completely disordered; individual monomer units comprising a chain do not have the motional independence associated with molecules in simple liquids. At the same time, the probability of defect formation in polymer crystals is essentially larger than in crystals consisting of more simple molecules. Since macromolecules are considerably longer than linear crystalline dimentions, individual chains may be followed through several crystalline and amorphous regions or can form loops and reenter the crystalline lamellae in random or as loose loops or regular, tight loops.

There is ample evidence that a spectrum of "degrees of order" exists in polymers that ranges between the perfect crystal and the perfect liquid. Associated with each degree of order is a structure that is intermediate between the extremes of crystal and liquid. Samples with the same density, which therefore have the same experimental crystallinity, may have completely different distributions of the various degrees of order. In this case, a single parameter is not sufficient to specify the physical composition of the sample. For a complete characterization, one must be able to describe all the structures and degrees of order present.

The picture of the amorphous phase in a semicrystalline polymer is rather vague. It is not clear if separate amorphous domains are present or the amorphous phase is located mainly at the surfaces of crystal lamellae or scattered in the form of defects and dislocations.

Polymer crystals are small (approximately 100–1000 Å) in linear dimensions, and the meaning of crystalline order over a range this small is unclear. The following question immediately arises: How much order may exist in a "noncrystalline" region before it is considered to be crystalline, or how small must a region of order be before it is no longer considered crystalline [46]? It appears that there is no unequivocal answer to these questions, since there is no absolute way to measure the degree of crystallinity.

Molecular interpretation of the multiple transitions in semicrystalline polymers is almost always more or less ambiguous. Even the assignment of such a large-scale transition as glass transition is in some cases uncertain. Polyethylene is a well-known example. A detailed description of the folded-chain crystal has to consider the unperturbed (i.e., ideal) crystal lattice which makes up the majority of the crystal, the chain folds at the main faces, lattice vacancies, interstitials, and eventually, the more or less ideal amorphous phase on the surface of the crystal. The chain folds may have the minimum length needed for a fold, may contain larger chain sections, or may represent mixtures of these possibilities. The chain end may protrude out of the fold-containing surface so that it remains somewhere on the surface outside the crystal lattice (cilia) and hence contributes to the "amorphous" phase at the crystal surface, or it may be located within the crystal. In the latter case, a point or a row lattice vacancy is created.

The glass-rubber relaxation (β relaxation) in semicrystalline polymers is in most cases complicated by the morphology. Two amorphous phases and, correspondingly, two glass-rubber transitions may exist, resulting from what is generally believed to be the chain-folded morphology of melt-crystallized material [25,47]. $T_g(L)$—the lower glass transition—is associated with dangling cilia and with molecules rejected by the spherulites. $T_g(U)$—the upper glass transition—is associated with loose loops or intercrystalline tie molecules.

Secondary relaxation in semicrystalline polymers is also complicated and may consist of two components, one originating in the amorphous component (γ_a), and the other arising in the crystalline phase (γ_c). The γ_c transition is believed to be associated with crystal defects wherein chains reorient over a low barrier.

Some polymers, especially those with chains having helical conformations, exhibit a relaxation process at temperatures lower than secondary relaxation. This relaxation is associated with the crystalline phase and is usually referred to as δ relaxation.

The complexity of the crystalline state also may be reflected in the appearance of an additional transition above the glass-transition temperature, the so-called α transition. A number of mechanisms have been suggested to explain the α crystal process. The theories can be divided into two classes: (1) those which treat the α process mainly as a property of the interior of the crystal, and (2) those which treat it mainly as a surface process. Representation of the α process by various authors as being an interior or, alternatively, a surface effect is not entirely misplaced—it very probably involves both [48].

Figure 6.7 shows schematically temperature transitions in a semicrystalline polymer and also, for comparison, the behavior of the polymer in the completely amorphous and completely crystalline states [48].

In most of the work relating mechanical behavior to morphology, comparison of various samples has been made on the basis of tan δ (the ratio of the imaginary part of the complex modulus to the real part) or some related quantity, such as the logarithmic decrement Δ. However, it is well known that the maximum value of tan δ against temperature is not a good measure of the strength of the relaxation unless the change in G', the real part of the complex modulus, is very small compared to the static modulus G_e [49]. This is not the case in most polymer systems. It is incorrect, therefore, to say that changing the degree of crystallinity increases or decreases the magnitude of a given relaxation process.

Other evidence [50,51] also indicates that the degree of crystallinity is not a sensitive enough criterion for sample comparison and that some of the relaxation processes are related to more subtle aspects of the morphology.

6.3.3 Individual Polymers

6.3.3.1 Polyethylene

Polyethylene as well as other crystalline polymers is not in thermodynamic equilibrium in the solid state. Depending strongly on the mode and conditions of sample preparation, it exhibits a variety of phase structures, one of which is a crystalline phase or region in which molecules or molecular chains are regularly aligned in space in a comparable manner to the crystals of monomeric substances. However, polyethylene is by no means fully crystalline—thermodynamic as well as kinetic considerations conclusively show the coexistence of a noncrystalline content, and it is thus revealed that polyethylene has a phase structure which includes at least two different regions with distinctly different molecular mobility. However, it has been generally recognized that the polyethylene rigid fraction evaluated by nuclear magnetic resonance is appreciably larger than the crystallinity observed by X-ray diffraction and density measurements. From a careful analysis of samples of different density and mode of crystallization, Bergman and Nawotki [52] have shown that the wide-line

MECHANICAL LOSS AT 1 Hz

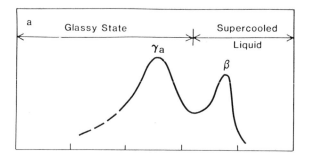

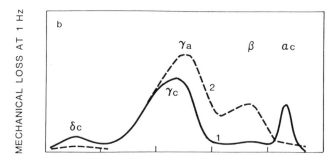

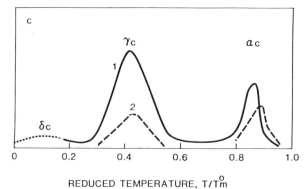

REDUCED TEMPERATURE, T/T_m^0

Fig. 6.7. Mechanical loss factor at 1 Hz versus reduced temperature T/T_m^0 for a linear polymer with no independently rotatable side groups: *(a)* amorphous state; *(b)* semicrystalline polymer (1 = high crystallinity; 2 = low crystallinity; and *(c)* single crystal mats (1 = high defect concentration and small period; 2 = low defect concentration and large period). After [48], © John Wiley & Sons, with permission. [Differently from the commonly used nomenclature (Section 4.2.1), the glass transition is labeled β in both semicrystalline and amorphous polymers.]

nuclear magnetic resonance spectrum of polyethylene consists of three superimposed bands having different widths, and a simple morphologic model can be developed to account for the foregoing nuclear magnetic resonance observations. In this model, the system consists of three zones, namely, (1) a crystalline zone with the chain units in an ordered conformation and with highly restrictive segmental motion, giving rise to a broad nuclear magnetic resonance band; (2) a disordered, amorphous interzonal region where the chain units are in a nonordered conformation and where segmental motion is relatively unrestricted, giving rise to a narrow band; and (3) an interfacial zone, many chain units thick, which is located at the diffuse boundary between the crystalline and disordered regions and in which the chain segments have an intermediate degree of mobility, giving rise to a band of medium width. The material in the interfacial and disordered zones constitutes the noncrystalline portion of the polymer. A model of this type has also been successful in explaining in a consistent manner over a wide range of molecular weights and crystallization conditions a variety of thermodynamic, spectroscopic, mechanical, and morphologic properties, as well as the very high interfacial free energy associated with high-molecular-weight samples of low density.

Similar to the nuclear magnetic resonance results, evaluation of the polyethylene structure by a Raman spectroscopy showed that in order to quantitatively describe the polyethylene spectrum it is necessary to invoke the existence of an anisotropic disordered region [53,54]. It was proposed that this region is located between the crystalline and amorphous layers and that in this region the chains are in an extended, trans configuration but have lost their lateral order. It was necessary to introduce this concept because the superposition of the reference spectra for the pure melt and pure crystal could not explain the observed spectra for semicrystalline samples in the CH_2 bending region.

The major characteristics of the Raman results follow the trends established by the nuclear magnetic resonance studies, and there thus appears to be a general concordance between the two methods. There are, however, essential differences between the two techniques when the data are examined in more detail. This reflects some of the underlying differences between the two methods and will be discussed later.

Low-Temperature Transitions in Polyethylene. Although perhaps the most extensively studied polymer, polyethylene is still the subject of controversy with respect to the nature of its thermal relaxation processes, especially the glass transition. The transitional phenomena of this polymer are the subject of an immense quantity of literature and several excellent reviews [25,55–57].

The temperature interval of particular interest is 100–300 K. Linear and branched polyethylenes exhibit three transitions at the following temperature regions: 145 ± 10, 195 ± 10, and 240 ± 20 K [25]. The methods which have been used to observe these relaxations cover the entire frequency scale from dilatometry to electron-sign resonance and nuclear magnetic resonance. The polyethylenes studied had different degrees of crystallinity, molecular weights, and morphology.

The lowest-temperature relaxation in polyethylene is the transition at 145 K, which has been termed the γ process. Some observations indicate that the same mechanism is responsible for the γ relaxation in crystals of n-alkanes, in single crystals of linear polyethylene, and in melt-crystallized polyethylene. The γ relaxations in polyethylene single crystals and in the n-alkanes both show the same dependence of temperature on the number of carbon atoms between the surfaces of the crystals [58]. The γ relaxation in melt-crystallized polyethylene also follows this relation, the number of carbon atoms being taken as that corresponding to the small-angle X-ray spacing [59]. Sinnott [58] suggested that the γ relaxation is associated with vacancies and/or dislocations within the lamellae. This mechanism does not necessarily imply that the relaxation is caused by reorientation of the actual vacancies and dislocations, since it could also be caused by reorientation of an unspecified molecular defect associated with these structural imperfections. Cooling to room temperature following formation of the crystal results in the defects being trapped within the crystal.

Various other mechanisms have been proposed for the γ relaxation observed in melt-crystallized polyethylene. It has been suggested that this relaxation is due to freezing-in of rotational isomers [59], to kinks in the chain [60], or to a local relaxation mode [61]. The latter requires thermally activated, hindered rotations around chemical bonds, and in terms of the barrier concept, these rotations involve the passage of atoms or atomic groups across a potential barrier from one position of equilibrium to another. The most widely known mechanism, the so-called crankshaft mechanism, was proposed by Schatzki [62] for polymers containing linear $(-CH_2-)_n$ sequences, where $n = 3$ or 4. This mechanism involves the simultaneous rotation about the colinear bonds such that the intervening carbon atoms move as a crankshaft. The requirement of the crankshaft mechanism is that the relaxation occurs in the amorphous region of the polymer owing to the condition that two bonds must be colinear. Such a condition is not fulfilled in crystalline regions where the $(-CH_2-)$ sequences are exclusively in the trans conformation.

The original suggestion, dating back to the first measurements of the dynamic-mechanical properties of polyethylene, was that the γ relaxation was due to motion of the polymer chains in the amorphous regions [63,64]. This suggestion was subsequently refined to motion in the amorphous regions of segments of the chain comprising three or four CH_2 units [65]. It was also proposed that γ relaxation has two components arising from amorphous (γ_a) and crystalline (γ_c) region contributions [48]. However, the well-established experimental fact that the γ relaxation in melt-crystallized polyethylene is due to the same mechanism as that in single crystals of polyethylene and in crystals of n-alkanes suggests that the γ relaxation in the bulk polymer does not arise from a discrete amorphous phase, but rather from within the lamellar structure. Although the experimental results do not negate the possibility of the existence of a discrete amorphous phase in melt-crystallized polyethylene, they do show that the existence of such a phase is not necessary to interpret the data.

In order of increasing temperature, the next transition in polyethylene lays at 195 ± 10 K and is the most controversial one in terms of its appearance and designation. Mechanical spectroscopy usually fails in resolving this transition, although Lam and Geil [66] succeeded in observing it by means of torsional-braid analysis in polyethelene ultraquenched in isopentane to the essentially amorphous state. On first heating, they observed, among others, the loss peak around 200 K followed by crystallization, as indicated by an increase in modulus. On a second heating, the peak at 200 K had been obscured by crystallinity. The other demonstrations of the transition around 200 K in polyethylene include Lee and Simha's data on linear thermal expansion [67] and Rathje and Ruland's low-angle X-ray scattering results [68]. In view of the extensive number of studies reported for linear and branched polyethylenes and the relatively few and, in some cases, tenuous observations of the transition around 200 K that have just been cited, it is clear that a definite pattern as to its existence in polyethylene has not as yet been established.

Following relaxation around 200 K, there is the transition at 245 K, and this is most readily registered by mechanical spectroscopy for highly branched polyethylene samples and polyethylene copolymers. Its intensity is related to the branching level and thus apparently to the degree of crystallinity. This transition either has not been observed at all in some linear polyethylene samples, has been barely detected in others, or has been found to be very sensitive to thermal history [69]. Pechhold *et al.* [70] and Illers [71] observed a definite shoulder in the 220–240 K temperature interval in torsion-pendulum experiments for bulk-crystallized linear polyethylene. Moore and Matsuoka [69] and Cooper and McCrum [72] also found a relaxation around 240 K in high-molecular-weight linear polyethylene samples in dynamic studies conducted at low frequency. The intensity of the transition was found to be very sensitive to the crystallization procedure.

This brief review of the transitions in polyethylene in the 100–300 K temperature interval makes abundantly clear that the structural basis and specific factors which govern these transitions are in need of further clarification. Each of the three transitions in 100–300 K temperature region has been labeled by more than one group of investigators as the true T_g (glass transition) of polyethylene. The literature is now formidable on this topic. None of these definitions, however, is markedly superior to the other. In a finely structured lamellar solid such as polyethylene, the T_g should not necessarily have all the physical characteristics of the T_g in completely amorphous polymers [15], since the amorphous forms of crystalline polymers are not in all respects identical to those in amorphous polymers. Since polyethylene cannot be obtained in a completely amorphous state, the question concerning its T_g value is somewhat academic. Recently, in an attempt to resolve a long-standing dispute in the literature about the glass transition in polyethylene, the concept of a double glass transition was proposed by Boyer [25] and subsequently extended to semicrystalline polymers in general [47].

In a series of papers published by Mandelkern *et al.*, it was concluded on the basis of the Raman spectroscopy [53,54] and dynamic mechanical [76] studies that the transition around 240 K in polyethylene results from the relaxation of chain units which are located in the interfacial region. Contrary to this conclusion, the nuclear magnetic resonance evaluations attribute this transition to liquid-like mobility in the interzonal region [73–75]. The major difference between the nuclear magnetic resonance and Raman results is that for low-molecular-weight bulk-crystallized polyethylene as well as for solution-crystallized polyethylene samples the former assigns all the noncrystalline contents to the interfacial region, whereas the latter attributes them to the interzonal region. From the morphologic standpoint, the nuclear magnetic resonance interpretation seems to be preferable, since it does not imply any restrictions to the generally accepted crystalline-state models, i.e., fringed micelle or any of the several variations of chain folding (with or without adjacent reentry). This is not the case with some of the Raman data, which cannot be explained if the regularly folded, adjacent-reentry type of interfacial structure is accepted. The major problem, however, of the assignment of the transition around 240 K to the interfacial region by Raman spectroscopy seems to be as follows: All the polyethylene samples evaluated by Raman spectroscopy exhibited from about three to eight times greater interzonal content when compared with the corresponding interfacial content [53,54]. Thus observation of the transition associated with the interfacial region seems to require the presence of the transition which manifests interzonal mobility. Since only transitions at 145 and 240 K are resolved by conventional mechanical spectroscopy, one has no other choice than to attribute the interzonal mobility to the transition around 145 K. The arguments against such an assignment are discussed elsewhere [25]. In addition, it has to be pointed out that the activation energy of molecular relaxation in the 130–150 K temperature interval does not exceed 10 kcal mol^{-1}, as estimated by both mechanical [70,187] and radiothermoluminescence [32] spectroscopies. This also indicates rather the local than cooperative nature of relaxation.

The difficulties mantioned above can be easily overcome by assignment of the transition in the vicinity of 240 K to the interzonal region. Then the absence of a transition associated with the interfacial region can be simply explained as due to the fact that it is insufficient to be resolved by mechanical spectroscopy.

Transitions Associated with Crystalline, Interfacial and Interzonal Regions as Evaluated by Radiothermoluminescence.   First experiments on the thermoluminescence of different polyethylene samples irradiated with γ-rays at liquid nitrogen temperature in the absence of oxygen showed that the transition in the vicinity of 140 K (the γ peak) was more intense relative to the remainder of the glow curve and much better resolved in high-density than in low-density polyethylene samples [36]. This was considered an indication that the γ relaxation in polyethylene originates in crystalline regions. The same conclusion has been made on the basis of results obtained for polyethylene samples allowed to stand at room temperature in hexane [77]. The effect of hexane on the polyethylene glow curve resulted in the disappearance of the transitions around 220 and 180 K, whereas the lowest-temperature transition at 130 K (γ transition) was only partially affected by hexane even after prolonged treatment. Partridge and Charlesby [41] found that polyethylene single crystals exhibited very intensive γ relaxation, whereas the luminescence intensity at higher temperatures was small relative to bulk-crystallized material. In a study of the radiothermoluminescence of highly crystalline linear alkanes [78] it was shown that such compounds exhibit a transition in the same temperature region as the γ transition in polyethylenes. Finally, the glow curves for both high-density and low-density polyethylene blown film samples showed a significant decrease in the intensity of the γ relaxation compared to that in bulk-crystallized samples, whereas the latter exhibited a higher degree of crystallinity [5].

The set of experimental facts cited above can be considered as evidence in favor of a crystalline-phase origin of γ relaxation in polyethylene. Contrary to the γ transition, complete removal of the transitions located at temperatures above the γ transition as a result of heptane treatment indicated that they originate in polyethylene amorphous regions [77].

As a rule, none of the conventional methods taken separately is sufficient to resolve all three transitions in polyethylene in the 100–300 K temperature region. However, the glow curves obtained by means of the radiothermoluminescence technique usually exhibit all three transitions. This fact supports Boyer's belief that temperature transitions within a composite semicrystalline polymer can best be seen only with some type of molecular probe [51]. Since trapping of electrons at low temperatures takes place in regions with structural imperfections, and since the subsequent thermoluminescence emission emanates from the same regions, one may consider radiothermoluminescence being kind of such a technique.

Figure 6.8 presents the radiothermoluminescence glow curves for two low-density polyethylenes [5]. In both cases, three low-temperature transitions are resolved. For both polymers, the temperature position of each low-temperature transition is the

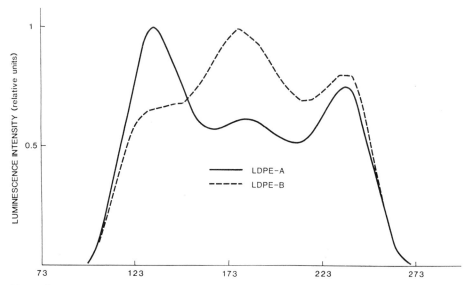

Fig. 6.8. Glow curves of two quenched low-density polyethylene samples. After [5], © John Wiley & Sons, with permission.

same, whereas the intensities are different. Since different polyethylene samples show temperature transitions located at the same temperature intervals, each of the transitions can be assumed to be derived from the mobility of a similar structural form. In this case, the intensity of a particular transition may be considered to be proportional to the mass fraction of a structural formation which is manifested by this transition.

In order to have a better understanding concerning the proper designation of each of the low-temperature transitions to mobility in specific structural regions, it is convenient to compare the data obtained by radiothermoluminescence analysis with the results obtained by other means.

A series of papers presented by Kitamaru and Horii [73–75] showed that three-component nuclear magnetic resonance spectrum analysis gives detailed information on the multiphase structure of polyethylene samples in terms of the molecular mobility or relaxation modes associated with each phase. It was concluded that bulk crystals generally have a structure composed of crystalline, interfacial, and interzonal materials. These interfacial and interzonal components are associated with limited and liquid-like molecular mobility, respectively. The relative content of these components, as well as the molecular conformation or mobility of each component, varies over a wide range depending strongly on molecular weight and temperature. The temperature dependence of the phase structure for annealed, monodisperse, linear polyethylenes of three different molecular weights as observed by Kitamaru and Horii [75] is shown in the middle section of Fig. 6.9. For an intermediate-molecular-weight sample at 120 K, the spectrum comprises only the crystalline component with no interfacial and interzonal components. The interfacial component is identified from the spectrum at 170 K in addition to the broad component. The narrow component is observable at temperatures above 230 K. These results suggest a large temperature dependence of the molecular mobility in each phase. The authors indicated that the

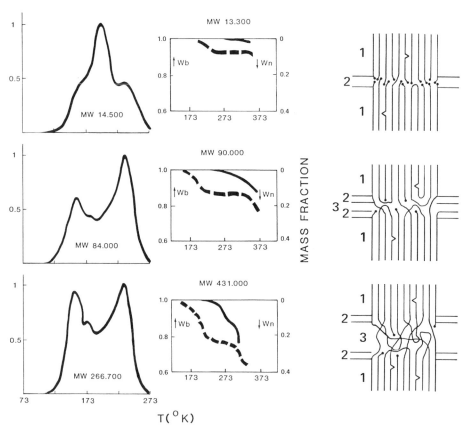

Fig. 6.9. *(Left)* Glow curves for linear polyethylenes. *(Center)* Mass fractions of three components mesured by the nuclear magnetic resonance technique as a function of temperature for linear polyethylenes. Solid and dashed lines indicate narrow and broad components, respectively. Wn, Wb = mass fractions of narrow and broad components, respectively (see ref. 75). *(Right)* Schematic structure models of different molecular weight linear polyethylenes: disheveled, unpeeled crystal for low-molecular-weight sample and lamellar crystals for high- and medium-molecular-weight samples. 1, 2, and 3 indicate the crystalline, interfacial, and interzonal regions, respectively. After [5], © John Wiley & Sons, with permission.

appearance of the interfacial component around 170 K and the interzonal component around 230 K corresponds to two low-temperature relaxation processes in polyethylene. A similar analysis has been made with samples of low and high molecular weights.

The left side of Fig. 6.9 presents the radiothermoluminescence results obtained for annealed monodisperse linear polyethylene samples with molecular weights similar to the samples analyzed by Kitamaru and Horii. The difference in molecular weights for high-molecular-weight samples is not important. The nuclear magnetic resonance technique has shown that samples of any molecular weight higher than 100.000 behave similarly to the sample with a molecular weight of 431.000.

Let us now compare the results obtained by both the nuclear magnetic resonance and radiothermoluminescence techniques. First of all, according to nuclear magnetic resonance, the appearence of interfacial and interzonal mobility shifts to lower temperatures with an increase in polyethylene molecular weight. There is a similar shift in the positions of the two thermoluminescence maxima. The highest-temperature transition shifts from 238 K for a low-molecular-weight sample to 232.5 for a high-molecular-weight sample. The corresponding shift for the intermediate-temperature transition is from 198 to 178 K. For samples of progressively higher molecular weights, the increase in intensity of the radiothermoluminescence maximum around 230 K also correlates with the increase in mass fraction of the interzonal component. Along with these similarities, there is one basic dissimilarity in the results obtained by the two techniques. According to nuclear magnetic resonance, the mass fraction of interfacial material increases with the increase in molecular weight, whereas radiothermoluminescence seems to show just the opposite effect, i.e., a decrease in combined intensity of the intermediate transitions located at 178 and 198 K. Let us consider this discrepancy in more detail. The spectrum of highly crystalline HNO_3-treated polyethylene at 123 K was taken as the elementary spectrum for the crystalline component. Although HNO_3 treatment probably destroyed the external noncrystalline portion of the material, it could not remove or heal the internal crystalline defects. So if some local molecular motion appears in the internal part of the crystalline region at temperatures above 123 K, its contribution would be assigned to the interfacial, but not crystalline component by the nuclear magnetic resonance technique. A somewhat similar situation has been noted by Sinnott [58], who observed a certain level of low-temperature mechanical relaxation in polyethylene samples having no nuclear magnetic resonance mobile fraction. Now, turning back to the radiothermoluminescence analysis, the higher the molecular weight, the greater is the intensity of the transition at 153 K (Fig. 6.9). This trend can be easily understood if one takes into account the increase in the degree of crystal imperfection with molecular weight. Thus the dissimilarity between the nuclear magnetic resonance and the radiothermoluminescence data is only an apparent one and can be explained by postulating that all three radiothermoluminescence transitions at 153, 178, and 198 K contribute to the interfacial nuclear magnetic resonance component.

The right side of Fig. 6.9 shows schematic structural models which correspond to low-, intermediate-, and high-molecular-weight polyethylene samples. These structural models are similar to those described by Kitamaru and Horii [75]. The only difference in our model is the presence of internal crystalline defects. The structure of a disheveled, unpeeled crystal is shown as the analog of the structure of a low-molecular-weight sample. Medium- and high-molecular-weight samples have lamellar crystal structures. The concentration of defects in the crystalline region increases with molecular weight. The low-molecular-weight sample has many chain ends and few loops and tie molecules in the interfacial region. The interzonal region consists predominantly of tie molecules and manifests itself to a greater extent in the samples of high molecular weight. Accepting this structural scheme, each of the radiothermoluminescence low-temperature transitions can be attributed to a specific structural region and to structural units in polyethylene; e.g., transition around 140 K, crystalline-region defects—kinks, jogs, etc.; transition at 178 K, interfacial region—cilia,

Table 6.2. Influence of different structural units in polyethylene on its mechanical properties

Property	Defects (kinks, jogs) crystalline region	(cilia, loose loops) interfacial region*	(chain ends, tight loops) interfacial region*	(tie molecules) interzonal region*
Tensile at yield	No influence	−	−	+
Elongation at yield	No influence	+	−	−
Impact strength	No influence	−	−	+
Environmental stress crack resistance	No influence	−	−	+

*Positive influence = +; negative influence = −.
Source: Ref. 5, © John Wiley & Sons, with permission.

loose loops; transition at 198 K, interfacial region—chain ends, tight loops; and transition at 233 K, interzonal region—tie molecules. Further, taking into account that the thermoluminescence intensity in the region of each particular transition is proportional to the mass fraction of the structural unit which is manifested by the transition, one can try to predict the mechanical properties of polyethylene on the basis of the radiothermoluminescence results. In order to achieve this goal, we must try to understand the influence of the different types of structural units on the mechanical properties of polyethylene. Such an estimation is to a large extent approximate; however, several conclusions are apparent (Table 6.2). For samples of the same degree of crystallinity, I would like to underline the positive influence of tie molecules and the negative influence of chain ends on the basic mechanical properties of polyethylene, e.g., yield strength, impact strength, and environmental stress crack resistance. However, crystalline defects seem not to have any influence, whereas an increase in cilia has a negative effect on yield strength and environmental stress crack resistance and a positive effect on elongation. Table 6.3 shows the basic mechanical properties for two low-density polyethylenes whose characteristics and glow curves are presented in Table 6.4 and Fig. 6.8, respectively. Sample A exhibits a higher intensity of the transition at 233 K (more correct, a larger area under the maximum at 233 K), which is indicative of a greater concentration of tie molecules. However, the intensity of the transition at 178 K is higher for sample B. Thus one can conclude that this sample has a larger cilia concentration. Since tie molecules predominately have a positive influence on the basic mechanical properties and cilia have a negative effect, one should expect better mechanical properties for sample A. The experimental results support this conclusion.

The Basic Difference Between Commercial High-Density and Low-Density Polyethylene. Type of Unsaturation and Its Influence on Transitions. The polyethylene molecule is composed of long sequences of CH_2 groups with alkylidene side branches of varied number and length. When polymerization is carried out at high pressure, the molecule may possess as many as 50 or more side branches, mainly ethyl and butyl, per 1000 carbon atoms. Heterogeneous catalysis carried out at low pressure using Zeigler or Phillips catalysts yields a molecule with far fewer predominantly methyl side branches, approximately 5 per 1000 carbon atoms. The polyethylene made at low pressure is generally more crystalline and therefore denser and stiffer

Table 6.3. Basic mechanical properties of two low-density polyethylene samples

Property	Sample A	Sample B
Tensile at yield (psi)	1730	1530
Elongation at yield (%)	18	31
Low temperature brittleness— WECO Notched (°C)	−41	+10
Environmental stress crack resistance —ASTM 10% Igepal (hs)	168	0.08

Source: Ref. 5, © John Wiley & Sons, with permission.

Table 6.4 Structural and molecular weight characteristics of commercial low- and high-density polyethylenes

Sample	Mw × 10^{-5}	Mw/Mn	Short-chain branches (GPS/100 Cs)	Unsaturation (GPS/1000 Cs)		
				Trans	Vinyl	Vinylidene
Low-density sample A	6.95	21.9	1.3	0.03	0.07	0.29
Low-density sample B	1.06	13.5	2.1	0.02	0.20	0.16
Low-density sample C	3.28	9.2	2.2	0.03	0.17	0.18
High-density sample 1	0.57	6.7	<1	<0.01	0.04	<0.01
High-density sample 2	0.45	8.5	<1	0.15	0.05	0.03
High-density sample 3	0.49	6.0	<1	<0.01	0.70	0.01

Source: Ref. 5, © John Wiley & Sons, with permission.

than the polyethylene made at high pressure. The former is often known as low-pressure, high-density, or linear, polyethylene to distinguish it from the latter, which is known as high-pressure, low-density, side-branched, or free-radical polyethylene.

It is well known that in solid polyethylene, the crystals have the form of twisted lamellae. Measurements of the unit-cell dimensions at 25°C have shown their increase with increasing numbers of side branches [79]. These results can mean either that methyl groups enter the lattice and so strain it internally or they do not enter the lattice but strain it externally. The question as to what extent the side branches can be accommodated within the crystal remains to be answered, although there are indications that short methyl branches can be included in crystalline domains [80]. It is certain, however, that bulky side branches (e.g., acetate groups) cannot exist in the crystal and that the major effect of side branches of any size is to increase the fraction of amorphous material.

Three forms of unsaturation are known to occur in polyethylene; these are vinyl (occurring at the ends of the main chain), trans (occurring in the middle of the main chain), and vinylidene (occurring in the side branches). Since the bond lengths for single and double bonds are different—0.077 nm for single and 0.067 nm for double bonds—it is possible that double bonds introduce some disturbance into the crystalline lattice which results in the formation of structural defects in their vicinity.

Low-density polyethylenes exhibit three low-temperature transitions on the glow curve located at 140, 178, and 233 K. The transition at 178 K associated with the interfacial region is the most intensive one for low-molecular-weight samples with high levels of short-chain branching (Fig. 6.8 and Table 6.4). The increase in the molecular weight results in the enhancement of the intensity of the transitions at 140 and 233 K affiliated with crystalline defects and tie molecules, respectively.

The main difference in the glow curves between commercial high-density and low-density polyethylenes is the absence of the transition at 178 K for high-density polyethylenes (Fig. 6.10 and Table 6.4). Only high-density polyethylene which is relatively rich in trans and vinylidene unsaturation showed a slightly pronounced shoulder at around 178 K, as well as the two transitions at 140 and 233 K. If one

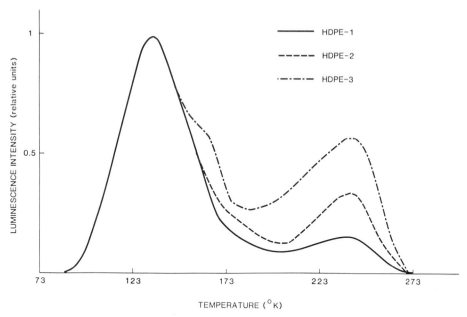

Fig. 6.10. Glow curves of three high-density polyethylenes. After [5], © John Wiley & Sons, with permission.

assumes that the high short-chain branching content in polyethylene is accompanied by high vinylidene unsaturation (at least for low-density polyethylenes this is the case), it is reasonable to conclude that this polymer has a relatively high, but still undetectable level of short-chain branching.

High-density polyethylene which is low in unsaturation shows only two transitions on the glow curve at 140 and 233 K.

The glow curve of high-density polyethylene which is rich in vinyl content has, in addition to the transitions at 140 and 233 K, a transition at 158 K which is pronounced as a shoulder. If one counts the total percent of unsaturation, this sample exhibits 5–10 times greater unsaturation than the other high-density polyethylenes. Such a large concentration of unsaturation should sharply increase the probability of large defect formation and simultaneously increase the number of cooperating subsystems which contribute to the relaxation in the interzonal region. Presumably, the first of these factors is responsible for the appearance of the transition at 158 K, whereas the second leads to the remarkable broadening of the transition at 233 K.

Annealed and Quenched Samples. Samples Obtained in the Process of Stress-Induced Crystallization.   Radiothermoluminescence has been found to be sensitive enough to reflect the changes in the polyethylene structure resulting from differences in crystallization conditions. Figure 6.11 presents the glow curves for a low-density polyethylene film-grade sample (sample C in Table 6.4) in which the samples were prepared under different conditions. Annealed samples were obtained by cooling 0.04-cm-thick compression-molded specimens (pressure 25.000 psi, temperature 160°C) in a

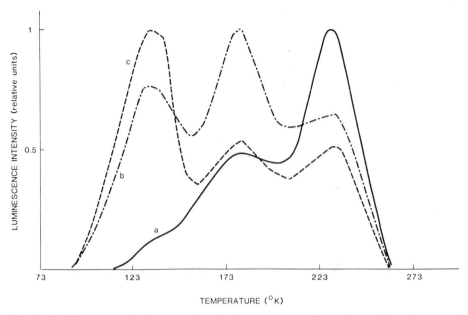

Fig. 6.11. Glow curves for low-density polyethylene sample C: *(a)* blown film; *(b)* annealed sample; and *(c)* quenched sample. After [5]; © John Wiley & Sons, with permission.

press to room temperature at a rate of 0.5 degrees/minute, whereas quenched samples were prepared by immersing a compression-molded specimen of the same thickness in ice water. As one can see, the blown film is characterized by the presence of all three low-temperature transitions typical for polyethylene, one of which (at 140 K) is very poorly pronounced. The transitions at 178 and 233 K coincide with the corresponding transitions in compression-molded samples. Differential scanning calorimetry analysis of blown film and compression-molded samples showed a higher melting point and lower heat of fusion for the blown-film sample. Generally, an increased value for the measured melting point can be attributed to several factors. They are either an increase in crystalline thickness, a decrease in the concentration of defects within the interior of the crystallites, or the orientation of the amorphous material caused by drawing. As is well known, stress-induced crystallization causes a decrease of the thickness of the lamellar crystal [81]. This eliminates the first possibility. The increase in melting point due to orientation of the amorphous material has to be accompanied by a simultaneous increase in the heat of fusion [82]. However, the heat of fusion has been found to be lower for the blown-film sample than for the compression-molded samples. The only remaining possibility (namely, the increase in the melting point due to a decrease in the concentration of defects within crystallites) correlates well with the very low intensity of the transition at 140 K for the polyethylene blown-film sample. The noticeable increase in intensity of the transition at 233 K reflects an increased concentration of tie molecules in the interzonal region. A three-region structural model is also helpful in explaining the differences in glow curves between quenched and annealed samples. In annealed samples, spherulites are larger and better organized. This accounts for the decrease in intensity of the transition at 140 K. Because of better organization, annealed samples must have fewer chains passing from one lamella to another. This causes a decrease in intensity of the transition at 233 K. At the same time, in the process of annealing, growing spherulites have "more time" to expel nonlinear defective material. This material concentrates on the spherulite surfaces and contributes to a relaxation in interfacial regions.

In contrast to low-density polyethylene, quenched and annealed high-density polyethylene samples prepared under conditions similar to those described above for low-density polyethylene did not show noticeable differences in the glow curves. This can be explained by a greater crystallization rate of high-density polyethylene in comparison to low-density polyethylene. The radiothermoluminescence results, however, indicated significant dissimilarity in the structure between compression-molded and blown-film high-density polyethylene samples. Basically, the changes are similar to those in low-density polyethylene. In the high-density polyethylene blown-film sample, the transition around 233 K becomes the most pronounced one, while the intensity of the transition around 140 K decreases. The most distinctive feature is the appearance of the transition at 178 K in the blown-film sample. In terms of the three-component structural model, it can be concluded that stress-induced crystallization is accompanied by the formation of an interfacial region. This region is probably generated at the expense of defective material which is pulled away from growing spherulites in the process of stress-induced crystallization. This explanation is in agreement with the conclusion that during polyethylene deformation single molecular-chain rearrangements lead to a chain segregation of defects into the amorphous phase [83].

The reason for the observed insensitivity of the shape of the high-density polyethylene glow curve to the preparation conditions (quenching versus annealing) is most probably due to the relatively thick samples and slow cooling rates utilized. When the thickness of high-density polyethylene samples was reduced to 0.0025 cm and the cooling rate was increased to 225 degrees/second, essentially different results were obtained [84]. The most significant findings can be summarized as follows:

First, in the temperature region of 200–250 K, the position, intensity, and half-width of the radiothermoluminescence peak depends on the cooling rate and decreases as the latter increases (Fig. 6.12). The authors indicated the consistency in the intensity and half-width of this peak for the samples A, B, and C. However, if one applies the peak deconvolution procedures to the experimental data in Fig. 6.12, certain changes in intensity and half-width become apparent.

Second, the intensity of the transition around 170 K increases and its resolution improves with increases in cooling rate, although its position remains the same.

Despite the decrease in half-width of the high-temperature peak with cooling rate, it is doubtful that complete amorphization has been achieved, since the half-width of this peak is still essentially larger than the 8–10 degrees characteristic of amorphous polymers [85]. The semicrystalline nature of ultraquenched samples also was confirmed by means of wide-angle X-ray diffraction. The diffractograms obtained at 120 K showed two sharp maxima at 2θ angles of 22 and 24.1°, indicating a degree of crystallinity of about 26% [86]. Even short-term room-temperature aging of the ultraquenched samples led to a change in the glow curves. In particular, the luminescence maximum situated in the 200–250 K temperature interval was displaced by 12–15°

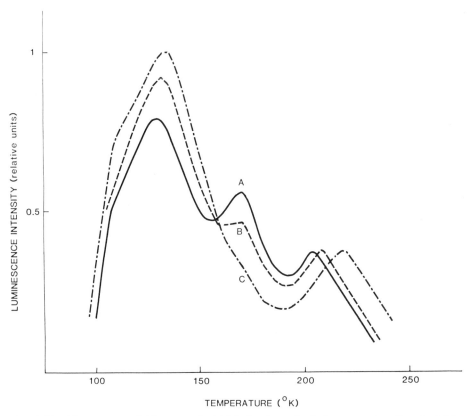

Fig. 6.12. Glow curves of thin, quenched high-density polyethylene samples. Sample thickness and cooling rate are, respectively, 0.0025 cm and 225 degrees/sec (A), 0.0025 cm and 60 degrees/sec (B), and 0.01 cm and 30 degrees/sec (C) [84].

toward higher temperatures and the degree of crystallinity of the sample kept for 30 min at room temperature increased up to 48%.

Thus quenching in liquid nitrogen of thin polyethylene films lowers the temperature position of the transition in the 200–250 K region, and this is apparently due to a decrease in the degree of crystallinity. The structure of the quenched samples is not stable and changes gradually during room-temperature aging. This is reflected in the increase in crystallinity and transition temperature.

The results presented above provide strong support for Boyer's double-glass-transition hypothesis [25], in which the "upper glass-transition temperature" $T_g(U)$ is postulated to be similar to the conventional glass-transition temperature $T_g(L)$ located at lower temperature but to involve motion of segments which are under strain. Let us remind ourselves that according to Boyer, $T_g(U)$ in polyethylenes lies around 240 K and $T_g(L)$ lies around 195 K. $T_g(U)$, as defined, must decrease with crystallinity and ultimately merge with the lower glass-transition temperature $T_g(L)$ of the completely amorphous polymer. Although such a merge was not observed because of incomplete amorphization, the trend toward it seems to be obvious.

Phillipov and Nikolskii's data on ultraquenched high-density polyethylene (Fig. 6.12) and the results on quenched low-density polyethylene disscussed earlier (Fig. 6.11) appear to be in contradiction in the sense that quenching of low-density polyethylene was found to be accompanied by a decrease in the intensity of the transition around 178 K and the opposite effect was noticed with high-density polyethylene. I believe that this is so because of much smaller differences in the degree of crystallinity between quenched and annealed low-density polyethylene, on the one hand, and quenched and ultraquenched high-density polyethylene samples, on the other. In the case of ultraquenched high-density polyethylene, an essential decrease in the degree of crystallinity overweights the largely defective nature of the small crystallites formed and the transition at 140 K decreases in intensity. At the same time, the number of strained tie molecules and the level of strain decreases and the portion of unstrained amorphous material increases, the overall effect being an increase in the intensity of the $T_g(L)$ transition accompanied by a decrease in the intensity of the $T_g(U)$ transition and its shift to lower temperatures.

Polyethylene Single Crystals and Crystals with Extended-Chain Conformations. Determination of the mechanism of the relaxations observed in polyethylene is severely hindered by the complex morphology of the melt-crystallized polymer. However, polyethylene solution-grown crystals as well as crystals obtained under high pressure have a relatively simple morphology which, although not completely understood, is nevertheless better characterized than that of the melt-crystallized polymer. Studies of molecular relaxation and its dependence on the supermolecular organization in polyethylene can best be carried out on samples produced from the same original batch, since the molecular relaxation depends on the crystallization conditions as well as on such material characteristics as the molecular-weight distribution, the concentration of substituents or branches, etc.

Extensive study of the relaxation behavior of Rigidex high-density polyethylene (Mw = 80.000) crystallized under different conditions has been carried out by Ozintszeva *et al.* [87]. The parameters for some of the samples studied are listed in Table 6.5, where samples A, B, C, D, and E represent the original material, single crystals, extended-chain crystals, samples with annular spherulites, and samples with radial

Table 6.5. Structural parameters of polyethylene samples

Sample	Crystallization temperature (°C)	Density (g·cm^{-3})	Crystallinity (%)	Melting point (°C)	Large period (Å)	$T_g(U)$ peak intensity (relative unit)
A	—	0.9565	70.5	132	120	1
B	77–78	0.9840	90	136	110	0.002
C	270	0.9940	95	144	None	0.1–0.2
D	132	0.9682	74.5	134	240	0.05–0.1
E	128	0.9686	74.6	134	240	0.05–0.1

Source: Ref. 87. Pergamon Journals, Inc., with permission.

spherulites, respectively. The details of sample preparation can be found in the original paper [87].

The glow curve of sample A (Fig. 6.13a) shows an intensive $T_g(U)$ peak in the range 200–240 K (the peak temperature equals to 223 K) as well as three peaks at lower temperatures, one at about 125 K and two less pronounced ones at 151 and

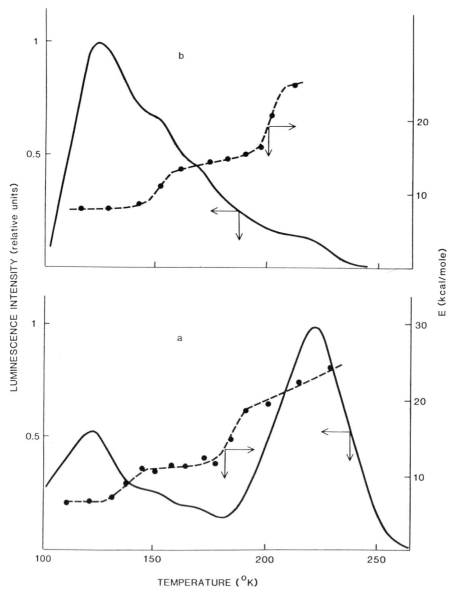

Fig. 6.13. Glow curves and activation energies of relaxation for high-density Rigidex polyethylene: *(a)* original sample A; and *(b)* single crystals B. After [87], Pergamon Journals, Inc., with permission.

171 K. The activation energies (E_a) in the ranges 100–135 and 145–170 K are 6.5 and 11 kcal mol^{-1}, respectively. E_a values, as measured in the 180–230 K temperature interval, increase from 15 to 25 kcal mol^{-1} with rising temperature. This, as well as the anomalous width of the $T_g(U)$ transition, points to a broad distribution of relaxation times.

The most intensive peak on the glow curve of sample B is situated around 125 K (Fig. 6.13b). The high-temperature descending part from this peak shows other less distinct peaks at about 150, 170, and 220 K. The determined value of E_a in the range 100–140 K did not exceed 7.5 kcal mol^{-1}. Starting from 140 K E_a rises, and at 170–220 K it is between 13 and 17 kcal mol^{-1}. The very weak $T_g(U)$ peak can be found in the range 200–230 K. E_a for this relaxation is 25 kcal mol^{-1}. A typical feature of sample B is its low thermoluminescence intensity throughout the temperature range studied. The glow curves in Fig. 6.13 been normalized per unit height of the most intensive maximum. To compare the luminescence intensities for different samples, it is necessary to use the Table 6.5, in which the $T_g(U)$ peak intensity of sample A is taken as the standard unit. Annealing of sample B at 120 and 132°C was accompanied by a 5- to 10-fold increase in thermoluminescence intensity, but the positions of the individual peaks and their relative intensities remained the same.

The most intensive maxima for sample C occur at 150 and 190 K (Fig. 6.14). Two less distinctive transitions can be noticed around 125 and 220 K. The activation energy of molecular relaxation smoothly increases as a function of temperature. This seems to indicate the broadening and overlapping of the relaxation spectra of various transitions. Annealing of samples crystallized under high pressure changed the form of the glow curve (Fig. 6.14b). When annealing was performed at 160°C and under normal pressure, it resulted in the appearance and growth of the maximum at 225 K as in the original polyethylene and in a noticeable decrease in the 150 K maximum; the maximum at 190 K proved more resistant to annealing.

Quite similar glow curves were obtained for samples D and E. Several peaks of approximately the same intensity at about 125, 148, and 167 K were noticed (Fig. 6.15), together with a relatively weak peak appearing in the range 210–215 K.

It follows from the results presented in Figs. 6.13–6.15 and Table 6.5 that changes in crystallization conditions are accompanied by essential structural changes, which, in turn, are reflected by radiothermoluminescence. One notices at the same time that the relaxation processes for all the samples have much in common, such as the number of the transitions and their temperature positions and approximately the same values of E_a. Especially noticeable is the similarity in radiothermoluminescence results for the samples D and E. This appears to be connected with the fact that temperature positions of local as well as cooperative molecular-relaxation processes depend mainly on the structure of the individual lamellae rather than on the dimensions or type of spherulites. The same conclusion seems to be valid for samples A, B, and C.

The temperature positions of various relaxations for samples A through E in the 140–250 K temperature interval coincide quite well with those discussed earlier. At the same time, all these samples showed the peak at 125 K. Although the position of this peak may be false, it indicates the presence of small-scale relaxation located

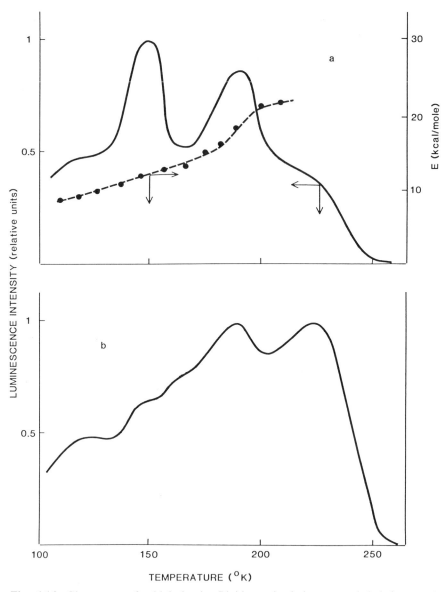

Fig. 6.14. Glow curves for high-density Rigidex polyethylene extended-chain crystals: *(a)* sample C; and *(b)* sample C annealed at 160°C. After [87], Pergamon Journals, Inc., with permission.

at $T \leqslant 125$ K. This same transition can also be found in some other polyethylene samples (Fig. 6.12, for example). In this context, it is worthy of note that the transition located at 115 K with an activation energy of 5.5 kcal mol^{-1} has been observed for high-density polyethylene irradiated at liquid helium temperature [88]. Its position on the temperature scale indicates that it is due to reorientation of small defects, such as of the type proposed by Reneker [89], within the lamellae. The

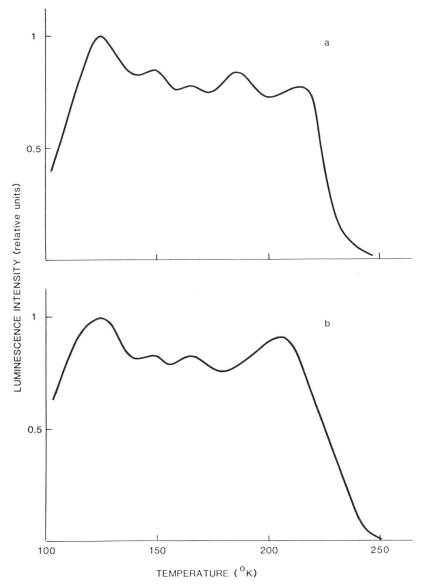

Fig. 6.15. Glow curves for high-density Rigidex polyethylene spherulitic samples: *(a)* sample D (annular spherulites); and *(b)* sample E (radial spherulites). After [87], Pergamon Journals, Inc., with permission.

specific defect proposed by Reneker is formed by compressing the planar zigzag chain in the direction of the chain axis so that one extra CH_2 group is incorporated in the compressed portion. Five CH_2 groups are then accommodated in the lattice in a distance along the chain axis normally occupied by four such groups. Reorientation of the Reneker defect, which is quite small and produces little distortion of the lattice from one equilibrium position to another, can be achieved easily by partial rotation of four carbon-carbon bonds. An estimated energy of formation of the Reneker defect is only 0.2 eV [58]. This value is close to the activation energy of 5–6 kcal mol^{-1} which was found to be the same for all the samples in the 110–140 K range. The practical coincidence of the Reneker-defect formation energy and the energy of molecular relaxation supports the proposed mechanism for the relaxation transition around 125 K. In particular, migration of the Reneker defects along the polymer chain at the annealing temperature is one of the mechanisms by which the single-crystal lamellae can increase their thickness [89]. This mechanism presumes the migration of defects along the length of the chain to the fold surface and the simultaneous injection of additional defects into the crystal to continue the process.

Consideration of the results reported by Osintzeva *et al.* [87] as well as the data on polyethylene relaxation discussed previously shows that the transition at 190 K is quite distinct only for extended-chain crystals, both low-molecular-weight samples crystallized under normal conditions and high-molecular-weight samples crystallized under pressure. This fact provides support for the designation of this transition as being due to relaxation of chain ends in the interfacial region. Since the extended-chain conformation is energetically more favorable than the folded-chain conformation, the transition around 190 K should in principle be very refractory to heating and disappear only as a result of melting and recrystallization.

The comparative evaluation of a variety of radiothermoluminescence experimental results with regard to relaxation transitions in polyethylene in the 120–250 K temperature interval provides an opportunity to attribute each of the transitions to a specific structural region and to certain structural units (Table 6.6). The assignment of the transitions at 233 and 178 K is in agreement with Boyer's double-glass-transition concept, where $T_g(U) = 233$ K and $T_g(L) = 178$ K [25]. Differently from Boyer, however, we attribute the transition at 140 K mostly to relaxations of vacancies and dislocations in crystalline regions, as was previously suggested by Sinnott [58]. Since prolonged heptane treatment of a variety of polyethylene samples influenced to some extent, although small, the radiothermoluminescence intensity of this transition [77], it may be of a composite nature, and a certain contribution from amorphous γ relax-

Table 6.6. Assignment of the transitions in polyethylene to certain structural regions and units

Transition temperature (K)	Structural region	Structural unit(s)
123 ± 5	Crystalline	Reneker-type defects
143 ± 5	Crystalline	Kinks, jogs
	Amorphous (interfacial, interzonal)	$(—CH_2—)_4$
178 ± 5	Amorphous (interfacial)	Cilia, loose loops
198 ± 5	Amorphous (interfacial)	Chain ends, tight loops
233 ± 10	Amorphous (interzonal)	Tie molecules

ation cannot be excluded. The transition at 198 K has been observed so far only by means of the radiothermoluminescence technique and only for polyethylene in extended-chain conformation. The transition around 120 K can probably be identified with Illers' γ''' crystalline transition [91].

Transitions in Polyethylene Above Room Temperature. In addition to the low-temperature transitions discussed above, polyethylene is known to exhibit at least one transition above room temperature but below the melting point. The nature of this α transition is less controversial than that for low-temperature transitions. It is generally agreed that it originates in the polyethylene crystalline phase. The best hypothesis at present for the mechanism of α relaxation is that it is due to vibrational or reorientational motion within the crystals [76].

As far as radiothermoluminescence is concerned, the observation of this transition(s) is seriously obscured by temperature quenching and by the fact that all charges stabilized at low irradiation temperature may well recombine before the onset of α relaxation. Nevertheless, several authors have observed thermoluminescence peaks between room temperature and the polyethylene melting point. For high-density polyethylene samples irradiated at room temperature, Blake *et al.* [92] reported the presence of three distinct regions on the glow curve with peaks at 51 and 72°C and a broad emission which dropped rapidly as the melting point was approached. A reduction in irradiation temperature resulted in attenuation of the peak at 72°C and weakening of the total glow observed. The peak at 50°C also was observed by Nakamura and Ieda [93] when irradiation was performed at room temperature in air. Hashimoto *et al.* [94] found four peaks at 50, 90, 120, and 140°C in polyethylene extended-chain crystals irradiated at dry-ice temperature. Several thermoluminescence peaks, although different in both shape and intensity from those in extended-chain crystals, were also observed in polyethylene single and folded-chain crystals.

It has to be underlined that the appearance of some high-temperature peaks on the glow curve may not have radiothermoluminescence origin. An example of this kind can be found in the paper by Nakamura and Ieda [93], who showed that the luminescence peak at 111°C from oxidized polyethylene samples can be seen irrespective of irradiation. This proves the chemiluminescence rather than radiothermoluminescence nature of emitted light. Thus caution should be excersized, and control experiments without irradiation should be carried out. It is certain, however, that at least some of the reported high-temperature thermoluminescence maxima in polyethylene can be observed only with preliminary irradiation [95]. Too little work has been done so far to allow any definite conclusions to be made with regard to the mechanism of high-temperature radiothermoluminescence and its relation to transitions in polyethylene.

6.3.3.2 Polypropylene

Tacticity, Morphology, and Transitions in Polypropylene. Polypropylene obtained by stereospecific catalysis is usually composed of three different steric configurations called *isotactic, atactic,* and *stereoblock.* Isotactic polymer is a linear polymer with a complete head-to-tail ordered arrangement and with the monomer units all disposed in the same steric configuration along the polymer chain, atactic polymer is a linear

polymer with a randomly ordered arrangement that does not possess a steric order of
the monomer units, and stereoblock polymer is a linear polymer with a partially
ordered arrangement and with a partial sequence of steric ordered units. Commercial
products usually consist of isotactic chains for the most part, but they also include a
minor fraction of atactic and stereoblock chains. The crystalline phase of polypropylene is composed of isotactic chains with a possible contribution from some long
isotactic sequences from stereoblocks. The amorphous phase, however, includes atactic
and isotactic chains as well as chain sequences furnished by stereoblocks, and consequently, it is probable that the nature of the amorphous phase depends on the ratio
of the three kinds of chain. Also, variations in tacticity can be expected to have
significant influence on polymer supermolecular structure, and indeed, this was found
to be the case. Miller [96] noticed that by quenching certain polypropylenes, a state
of organization could be reproducibly obtained which was neither amorphous nor
crystalline. To distinguish this form of organization from crystalline and amorphous
states, he called it the ''noncrystalline'' form. Perhaps the most graphic illustration
of this different form of polypropylene is given by its X-ray scattering (Fig. 6.16).
Although the scattering curve of the quenched sample is similar to that of the atactic
sample, the presence of a second scattering maximum at about 22.2 degrees indicates
the existence of a state of aggregation or order greater than that attained in the atactic
sample, the latter being considered to be completely amorphous. However, the absence of any of the crystalline diffraction maxima exhibited by the annealed (crystallized) sample indicates that the extent of perfection of this aggregation is not sufficient to cause X-ray diffraction. Since a crystal as small as 50 Å (and perhaps even

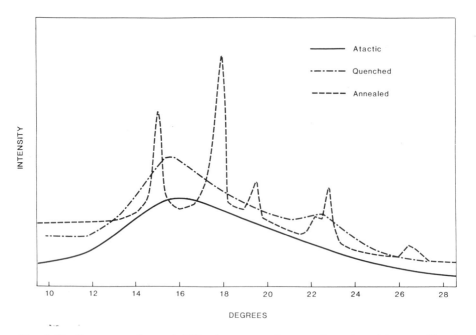

Fig. 6.16. X-ray scattering and diffraction curves for amorphous (atactic), crystalline (annealed), and noncrystalline (quenched) polypropylenes. After [96], IPC Science and Technology Press Ltd., with permission.

smaller) can be detected by its diffraction, it can be concluded that the quenched sample exists in a state of organization intermediate between those of the amorphous and crystalline states. This form was characterized as being an aggregation of molecules (or segments of molecules) in which portions of the individual chains maintain the helical structure found in the crystal, but in which there is little or no lateral order sufficient to be called crystalline. The order existing in the noncrystalline form might be that described by Hosemann [97] as "paracrystalline," which is defined as a deformation of the crystal structure obtained by replacing the constant cell edges by a statistically determined vector varying in both length and direction, or alternatively, as described by Natta [98] as "unstable smectic," meaning that right- and left-handed helices are arranged randomly to give a pseudohexagonal structure, whereas in the normal monoclinic unit cell the helices are regularly arranged with respect to one another.

As opposed to polyethylene, which exhibits transitions in well-defined temperature regions, there are in many cases striking differences in the reported polypropylene transition temperatures even when the studies were carried out by means of the same technique [99]. The only transition which was reported by practically all researchers is the glass transition of amorphous polypropylene that lies slightly below 273 K, although it is uncertain whether there is a difference between the glass transitions in atactic and isotactic materials [100,101]. Different groups of authors found none or several transitions in the range 173–273 K, and in some cases, transitions also were observed at lower (down to 19 K) and higher (up to 373 K) temperatures. Such behavior is indicative of the scatter nature of the amorphous phase in polypropylene.

In order to explain the peculiarities of relaxation in polypropylene, a concept of a distributed transition temperature was introduced [102]. According to this concept, the appearance of the transition is connected not with changes in the macromolecules themselves, but with a different kind of movement in the intermolecular bonds. The existence of different supermolecular formations with bonds between them results in a "distribution" of the transition temperature over a very wide range of temperatures. This distribution may be uniform or quite otherwise, depending on features of the polymer structure. Thus a temperature transition determined, for instance, by mechanical spectroscopy may be expressed either as one maximum or as a limited number of maxima, or as a plateau. Although the distributed-transition-temperature concept is helpful in explaining some of the experimental results, it neither relates transitions to any concrete form of molecular motion nor predicts variations in relaxation with supermolecular structure.

I believe that the essential dissimilarities in polypropylene transition temperatures reported in the literature are due primarily to a low level of isotacticity and thus to the high atactic and stereoblock content in some of the polymers introduced as isotactic. The ambiguity of this situation may stem in part from a failure to study polymer samples that have been carefully characterized with respect to both molecular weight and tacticity and from a lack of consistency in measurement. The use of a high-resolution technique seems to be necessary. Careful tacticity determination is crucial, since the steric purity of polypropylene is rarely complete, being determined by catalyst type as well as reaction conditions. Solvent extraction or infrared measurements cannot be regarded as a reliable standard for tacticity determination, since these methods are dependent on the molecular weight and crystallinity of the polymer samples [101]. Consequently, an absolute method for tacticity determination, such as ^{13}C nuclear magnetic resonance, must be considered mandatory for establishment of the stereochemical configuration. Also, molecular-weight evaluation is important, because it was shown that the T_g value for atactic polypropylenes falls dramatically in the low-molecular-weight region [103].

Heptane-Soluble and -Insoluble Fractions. Atactic Polypropylene. Properties of polypropylene are to a great extent dependent on tacticity and the amount of low-molecular-weight soluble material in the polymer. Fractional extraction with different solvents in a Soxhlet-type apparatus is widely used in industry for the preliminary screening and evaluation of material quality.

Important factors when considering the dissolution of semicrystalline polymers are solvent temperature, polymer density, and molecular weight [104]. At temperatures well below their melting points, semicrystalline polymers do not completely dissolve in organic solvents. The positive entropy changes associated with fusion and mixing do not compensate sufficiently for the large heat of fusion, so overall, the free-energy change for crystallite dissolution is positive. As mentioned, density is another important factor. The dissolution rate of polypropylene has been observed to increase with decreasing polymer density, although not in a simple manner [105]. Finally, one must consider the polymer molecular weight. The entropy per unit volume of a solution at constant volume or weight fraction decreases with increasing size of the solute molecules. Free energy will thus increase and solution stability will decrease, with the result that increasing the molecular weight of the polymer makes dissolution more difficult [106].

Semicrystalline polymers such as polypropylene can be considered to be composite materials, consisting of purely amorphous and purely crystalline regions, with a wide morphologic spectrum in between. It is expected that the dissolution behavior of such a polymer is complex, with contributions from each region. Above T_s, the thermodynamic dissolution temperature, all fractions of the polymer are thermodynamically soluble, but in a short time, amorphous and low-molecular-weight crystalline fractions are kinetically preferred.

Glow curves for three polypropylene samples with the same heptane-insoluble content (89.5%) are shown in Fig. 6.17. Despite the same level of heptane insolubles, the differences in shape of the glow curves indicate significant structural differences between the samples. At the same time, one notices that the temperature intervals where the intensity of emitted light exhibits maxima are similar for all the samples, namely, 273–263, 223–213, and 143–133 K. In one case (curve *c*), a poorly pronounced transition around 173 K also can be noticed. The radiothermoluminescence evaluation of a large number of polypropylene samples produced by different manufacturers showed that these same transitions are characteristic for all polypropylenes having a relatively high level of isotacticity [107].

The so-called atactic fraction obtained from the customary heptane extraction of polypropylene actually may consist of a series of various components in variable weight ratios which are similar only in their solubility in heptane: noncrystallizable atactic polypropylene, crystallizable relatively low-molecular-weight stereoblock

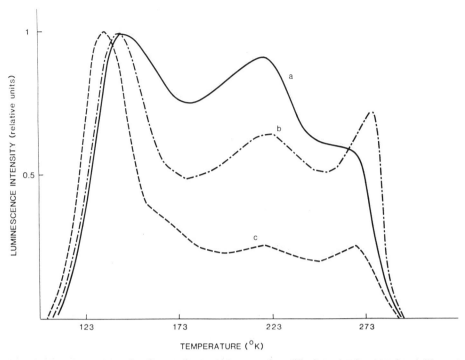

Fig. 6.17. Glow curves for three polypropylene samples with the same heptane-insoluble content [107].

polypropylene, and extremely low-molecular-weight isotactic polypropylene. Thus both heptane-soluble and -insoluble fractions can be expected to contain various components depending on the actual composition of the material before extraction. This aspect has been underlined by several authors [108,109], and it was emphasized that the evaluations based on material balance only may be misleading. The radiothermoluminescence results support this conclusion. Figure 6.18 presents the glow curves for two virgin polypropylene samples (A and B) and their heptane-soluble and -insoluble fractions. The major difference between samples A and B is the greater intensity of the broad transition centered at 223 K and the higher temperature of the transition in the vicinity of 273 K for sample A. At the same time, the dissimilarities in the glow curves of the heptane-soluble and -insoluble fractions of these materials are dramatic. The soluble fraction of sample B most probably represents completely atactic polypropylene with a narrow maximum in the T_g region and essentially broader secondary relaxation transition at 213 K. Since the lowest temperature transition for amorphous polypropylene is located at 213 K, the transitions at 133 and 163 K must arise from relaxation in polypropylene crystalline regions unless the relaxation behavior of the amorphous phase formed by atactic and isotactic chains differs essentially. Following this line, it can be concluded that the soluble fraction of sample A contains some crystallizable material. This results in increased width of the maximum around T_g and its shift by 8 degrees to higher temperatures as compared to sample B, and it

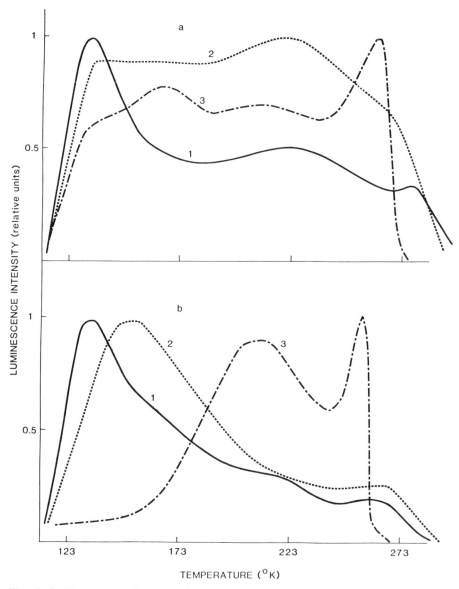

Fig. 6.18. Glow curves for two polypropylene samples: *(a)* sample A; and *(b)* sample B (1 = virgin materials; 2 = heptane-insoluble fractions; 3 = heptane-soluble fractions [107].

also results in the appearance of the two low-temperature transitions at 133 and 163 K. The characteristic features of the insoluble portion of sample A are an intensive broad transition at 223 K, the appearance of a very weak transition at 163 K, and a poor resolution of the thermoluminescence maximum in the region of the glass transition. The major difference between virgin sample B and its heptane-insoluble fraction is in the position and shape of the most intensive lowest-temperature maximum. For the heptane-insoluble fraction, the temperature position of this maximum (150 K), as well as its large width, indicates that it may be composed of two overlapping transitions of about the same intensity located at 163 and 133 K.

The limited radiothermoluminescence results on polypropylene relaxation obtained so far do not permit a substantiated assignment of different transitions to specific structural formations. It has to be underlined, however, that as with polyethylene, relaxation in polypropylene manifests itself in certain temperature intervals. An additional indication of this was obtained by evaluation of highly isotactic polypropylene samples of different molecular weights [107]. The same four transitions as mentioned earlier were observed. The intensity of the transitions around 133 and 263 K increased with molecular weight, whereas two other transitions at 163 and 223 K were barely pronounced, especially in low-molecular-weight samples.

The only definite conclusion which can be made at the present time is that completely amorphous atactic polypropylene exhibits glass relaxation and secondary relaxation around 260 and 220 K, respectively. It is still unclear, however, whether the amorphous phase in isotactic polypropylene can be identified with that in atactic material. The transitions at 163 and 133 K most probably are caused by the relaxation of different defective sites in the polypropylene crystalline phase.

It is noteworthy to mention that, as a rule, commercially available atactic polypropylenes are not completely amorphous and exhibit both amorphous and crystalline transitions [90]. Thus any studies on polypropylene relaxation have to be performed on carefully and thoroughly characterized samples.

Crystallization Conditions and Molecular Relaxation in Polypropylene. Elucidation of the relaxation behavior of the same material crystallized at different conditions is a usual route taken in an attempt to establish the relationship between polymer structure and molecular relaxation. This kind of study has been carried out with the commercial isotactic polypropylene Moplan ($Mw = 2 \times 10^5$) [110]. The original material (layers 0.3 mm thick cut from polypropylene granules) exhibits a very intensive and broad thermoluminescence maximum in the temperature range of the glass transition (Fig. 6.19a). The glass-transition temperature T_g of this sample determined from the position of this maximum is 256 K, which practically coincides with the T_g value for completely amorphous atactic material, although the half-width of the T_g peak is six to seven times larger in the former case. Two other weak maxima are located at 125 and 163 K and correspond to the crystalline transitions in polypropylene. It is not clear whether the transition around 223 K noted for other isotactic polypropylenes is absent or masked by the low-temperature part of the peak at 256 K. In this respect, it is of interest to note that the activation energy of molecular relaxation rises sharply just in the 213–223 K region and then shows a more gradual increase. This seems to indicate the presence of molecular motion with a relatively narrow distribution of relaxation times around 223 K and a broader distribution at higher temperatures. It has to be underlined that the overall shape of the glow curve for the original polypropylene Moplan differs essentially from those for conventional isotactic materials: the T_g peak is dominant and shifted by 10–15 degrees to lower temperatures, and the

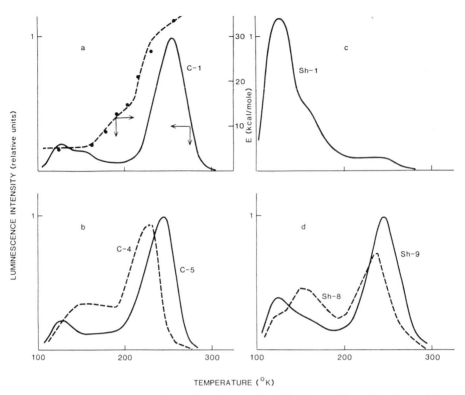

Fig. 6.19. Glow curves for the isotactic polypropylene Moplan: *(a)* the original material, C-1 *(b)* remelted samples C-4 and C-5; *(c)* "shish kebab" crystals obtained by crystallization from 1% solution in tetrachloroethylene at 85–90°C; and *(d)* "shish kebab" samples Sh-8 and Sh-9 [110].

transitions below 173 K are very weak. Unfortunately, data on tacticity were not provided by the authors [110]. Thus a parallel comparison on the basis of stereo-chemical composition cannot be made.

Variations in the conditions of polypropylene annealing markedly affect the positions of individual transitions and their intensities (Table 6.7 and Fig. 6.19*b*). The high-temperature annealing or remelting of C-1 granules noticeably reduces T_g. This is most clearly observed in polypropylene samples which had been slowly cooled (C-2 and C-4). It appears at the same time that even for quenched samples, T_g is lower by 10–15 degrees than the T_g for sample C-1.

As with polyethylenes crystallized from a dilute solution, the Sh-1 sample exhibits the most intensive maximum at low temperatures (129 K); two other transitions around 170 and 245 K are poorly pronounced (Fig. 6.19*c*). Annealing at temperatures above 135°C but below 180°C and subsequent slow cooling to room temperature improve resolution of the maximum in the T_g region and shift it to lower temperatures. The positions of the two low-temperature maxima remain unchanged, although the intensity of the transition at 129 K slightly decreases with the annealing temperature,

Table 6.7. Influence of annealing on relaxation transitions in the polypropylene Moplan

Sample	Preparation conditions	Transition temperature (K)	
		Secondary transition	Glass transition
C-1	Original material	160*	256
C-2	Annealing at 160°C for 4 hours, slow cooling (1 degree/minute) to room temperature	170	236
C-3	Annealing at 160°C for 4 hours, rapid cooling (30 degrees/second) in a water-ice mixture	170	243
C-4	Heating to 190°C in 0.5 hour, slow cooling (1 degree/minute) to room temperature	160	232
C-5	Heating to 190°C in 0.5 hour, rapid cooling (30 degrees/second) in a water-ice mixture	Not observed	245
C-6	Heating to 190°C, crystallization at 134°C for 15 min, cooling to room temperature (30 degrees/minute)	159	245
C-7	Heating to 190°C, crystallization at 134°C for 4 hours, cooling to room temperature (30 degrees/minute)	170	245
Sh-1	Crystallization from 1% solution in tetrachloroethylene at 85–90°C	170*	242*
Sh-2	Annealing of Sh-1 for 4 hours at 135°C, slow cooling (1 degree/minute) to room temperature	170*	215*
Sh-3	Annealing of Sh-1 for 4 hours at 150°C, slow cooling (1 degree/minute) to room temperature	170*	215*
Sh-4	Annealing of Sh-1 for 4 hours at 160°C, slow cooling (1 degree/minute) to room temperature	169	215*
Sh-5	Annealing of Sh-1 for 4 hours at 168°C, slow cooling (1 degree/minute) to room temperature	170	210
Sh-6	Annealing of Sh-1 for 4 hours at 168°C, rapid cooling (30 degrees/minute) in a water-ice mixture	172*	245
Sh-7	Annealing of Sh-1 for 4 hours at 175°C, slow cooling (1 degree/minute) to room temperature	171	211
Sh-8	Annealing of Sh-1 for 4 hours at 190°C, slow cooling (1 degree/minute) to room temperature	160	236
Sh-9	Annealing of Sh-1 for 4 hours at 190°C, rapid cooling (30 degrees/second) in a water-ice mixture	Not observed	245

*Very weak transition, of which the position is recorded with an accuracy of $\pm 3°$.
Source: Ref. 110, Pergamon Journals, Inc., with permission.

whereas the intensity of the transition at 170 K shows the opposite tendency. Further rises in the annealing temperature bring about marked structural changes. Samples Sh-8 and Sh-9 heated to temperatures exceeding 182°C, the melting point of sample Sh-1, exhibit glow curves of the same shape as those of samples C-4 and C-5, respectively (Fig. 6.19d), and differ from them only in the intensity of emitted light. This similarity reflects the transformation of a fibrillar morphology characteristic of solution-crystallized samples into a lamellar morphology conventional for melt-crystallized samples. Like samples of series C, annealed and then quenched samples of series Sh have a higher T_g than corresponding samples slowly cooled after annealing.

Along with high-temperature annealing, room-temperature aging also was found to affect polypropylene relaxation. Room-temperature aging of both quenched and slowly cooled samples is accompanied by a gradual increase in T_g. The glow curves in Fig. 6.19 for samples C-4 and C-5 were plotted 20–30 hours after sample preparation. On the glow curves of the same samples plotted 20–40 days later, the T_g maximum was displaced by 5–10 degrees in the direction of high temperatures. A similar effect was observed when samples of series C were further annealed for several hours at 80–90°C immediately after the main heat treatment. The responses of samples Sh-8 and Sh-9 to room-temperature aging were identical to those for samples C-4 and C-5.

The effect of high-temperature annealing and room-temperature aging on the position of the relaxation transitions in polypropylene has been repeatedly noted previously. A gradual increase in T_g of isotactic polypropylene during storage at room temperature has been observed and examined in detail [111,112]. The latter effect is usually explained by the fact that the amorphous chains are fairly rigidly fixed and are subjected to high constraint by surrounding crystallites. Gradual structural ordering during secondary crystallization forces the amorphous chains into a more strained state, so T_g is increased.

It has also been indicated by many authors [96,113–115] that the slow cooling of polypropylene from the melt lowers T_g by a few degrees. Assuming that crystallization of isotactic polypropylene is accompanied by enrichment of amorphous regions by an atactic polymer, low-molecular-weight impurities, etc., Passaglia and Martin [115] suggested that displacement of the atactic fraction into the interlamellar space is the cause of the reduced T_g. If this is the case, one would expect a marked difference in T_g between the samples of series C and Sh, since during crystallization from a dilute solution, fractionation must take place to a much greater extent than during crystallization from the melt. However, similar relaxation characteristics of samples Sh-8 and Sh-9 on the one hand and samples C-4 and C-5 on the other indicate that fractionation is not the main reason for variations in T_g.

It is therefore more likely that the stresses arising in amorphous regions between adjacent lamellae according to conditions of crystallization, annealing, and cooling should be regarded as the cause of variations in T_g. Crystallization is accompanied by an increase in the density of the polymer, whereas the crystallites already formed reinforce polypropylene and impede shrinkage. As a result, forces arise that stretch the amorphous, disordered polymer segments, and this may be accompanied by molecular orientation as well as micropore and crack formation. Accepting this scheme, the main relations governing variations in T_g can be explained. For example, the low

T_g values of samples slowly cooled from melt may be due to the fact that, under these conditions, the contact between adjacent lamellae weakens, the most stressed tie molecules relax in the 100–160°C region as a result of mobility inside the lamellae, while microcrack formation reduces the conformational dimensions of the relaxing subsystem and thus facilitates the glass relaxation. The reduction in conformational dimensions of the relaxing subsystem due to cracking may progress to such an extent that the system is not large enough for complex relaxations requiring certain cooperations between adjacent macromolecules. Then the glass relaxation vanishes, and the regions with insufficiently large volume and a disordered arrangement of macromolecules contribute to secondary local-mode relaxation. The polypropylene samples crystallized from a dilute solution have more regular lamellar surfaces and less interlamellar contacts and may be expected to exhibit this effect more readily.

The region 213–223 K seems to be the low-temperature limit of the polypropylene glass transition (Table 6.7). Since the secondary relaxation in atactic material was found in the same temperature interval, it is possible that the transition around 215 K in samples of the Sh series manifests secondary rather than glass relaxation.

Effect of Shock-Wave Compression on Polypropylene Molecular Relaxation. A *shock wave* is a pulse of pressure that creates one-dimensional compression of a substance and moves through it at supersonic speed with a pressure amplitude exceeding the ultimate strength of the material [116]. The passage of a shock front through solids causes intensive microplastic deformations. Shock compression occurs within 10^{-6} sec and is not isentropic, whereas the relief of pressure by rarefaction waves is an isentropic process which occurs in less than 10^{-5} sec, causing rather high residual temperatures. Shock compression performed at liquid nitrogen temperature eliminates the effect of high residual temperatures and fixes the shock-induced structural changes.

Shock compression with pressures up to 400 kbar carried out at 77 K does not change either morphology or the structure of supermolecular formations in polypropylene [117]. At the same time, examination of samples by means of radiothermoluminescence showed rather large changes in the relaxation spectrum. These changes are especially pronounced in the case of the oriented polypropylene samples (Fig. 6.20). The glow curve of the original oriented sample is similar to that of the same material (Moplan) crystallized from a dilute solution (Fig. 6.19c) and reflects the peculiarities of a fibrillar structure. As a result of shock compression at 77 K, the transition in the T_g region vanishes and, simultaneously, a well-defined maximum appears at 163 K. Thus low-temperature shock compression is accompanied by a dissipation of polypropylene glass relaxation which most probably arises from the formation of numerous cracks in the amorphous phase, providing ''loosing'' in this phase. The maximum at about 163 K seems likely to be associated with relaxation on the surfaces of the cracks. Heating of the shock-compressed sample to room temperature lowers the intensity of the transition at 163 K and restores the cooperative relaxation, although it appears at somewhat lower temperatures, indicating incomplete healing of the cracks.

Since one would expect the presence of a transition associated with secondary relaxation in a semicrystalline material whose glass relaxation has vanished, the ab-

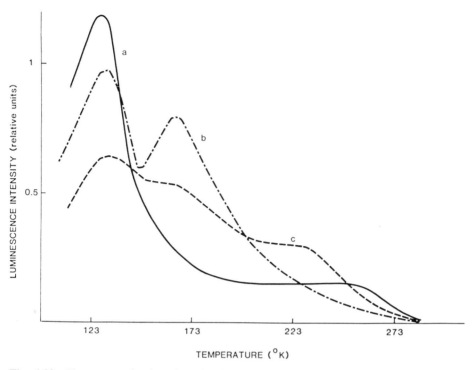

Fig. 6.20. Glow curves for the oriented polypropylene Moplan: *(a)* original sample; *(b)* sample shock-compressed at 77 K; and *(c)* sample *b* heated up to 293 K. After [117], © John Wiley & Sons, with permission.

sence of a transition around 223 K brings some ambiguity to the previous assignment of the transition at 163 K as originating in the polypropylene crystalline phase. It can be argued that this represents the secondary relaxation of the mainly isotactic amorphous phase of polypropylene. Nevertheless, the microcracks are most probably formed in both the crystalline and amorphous phases, and from this point of view, assignment of the 163 K transition as being due to the relaxation of large defects in the polypropylene crystalline phase remains valid.

The very weak glass relaxation in the original material and the absence of any amorphous transitions in the shock-compressed sample suggest a very highly defective crystal or paracrystal in the as-drawn state, in which crystalline and amorphous regions are not clearly delineated. A great reduction in segmental material in the amorphous regions as a result of polypropylene drawing with no loss peak in the glass-transition region has been observed previously by means of the mechanical spectroscopy [118].

It is to be noted that the compression of polypropylene samples with pressures up to 25 kbar at 77 K under static conditions does not cause any change in the shape of the glow curve. Also of interest in this connection are the results of mechanical tests of isotropic polypropylene samples after shock compression at −196°C followed by warming to room temperature [117]. Although microscopic examination did not show any morphologic or structural differences between the original and the shock-compressed samples, marked dissimilarities in their ability to deform under load were observed. When the original isotropic polypropylene sample was subjected to uniaxial deformation at room temperature, all three regions of the stress–strain curve were found, with an elongation at break of as much as 500–600% [117]. For shock-compressed samples, however, only the initial "Hookean" region in the curve was observed, with fracture occurring at 15–20% elongation. The first region of the stress–strain curve—prior to necking—is known to be due mostly to polymer deformation in the amorphous regions between crystallites [119]. The existence of this region in the shock-compressed samples points out that, upon heating to room temperature, the concentration of defects generated by shock-wave action decreases sharply in the amorphous regions. At the same time, the fact that brittle fracture of these samples occurs prior to yielding indicates that defects situated in the crystalline regions are thermally annealed to a relatively lesser extent than those in amorphous regions. These defects are likely to act as stress concentrators during the deformation, leading to the initiation and propagation of large cracks and finally to rapid fracture of the test sample.

6.3.3.3 Polytetrafluoroethylene

Polytetrafluoroethylene is a nonpolar polymer with essentially linear molecules containing two fluorine substituents on each main-chain carbon atom. Polytetrafluoroethylene as polymerized has a very high degree of crystallinity and nearly perfect three-dimensional order. Unlike most crystalline polymers, polytetrafluoroethylene, as ordinarily prepared, is not spherulitic, although it is suggested that the chains in polytetrafluoroethylene structural formations are folded as in the other crystalline polymers [120].

All researchers reported the presence of two (in some cases one) peaks near room temperature on the glow curve of polytetrafluoroethylene. Mele *et al.* [121] observed a broad maximum at 308 K for material irradiated in a vacuum with a very low dose and two maxima at 292 and 327 K at higher doses (~0.1 Mrad). Nikolskii [10] noticed two maxima at 269 and 297 K, and Tomita [122] noticed two at 265 and 295 K. These transitions most probably correspond to the "crystal-disordering" transitions at 292 and 303 K first observed by discontinuous changes in density [123] and later by specific-heat maxima [124], and consequently, they can be considered first-order in the thermodynamic sense. X-ray diffraction [125] and nuclear magnetic resonance [126] also show crystal–crystal transition in the vicinity of 293 K. At 292 K, unit-cell changes occur from triclinic to hexagonnal. At 303 K, the preferred crystallographic direction is lost and the molecular segments oscillate about their long axes with a random angular orientation in the lattice. Attribution of these two transitions to crystalline-phase rearrangements correlates with the observation that the greater is crystallinity, the higher is the intensity of thermoluminescence [10].

There is less agreement, however, on the position of the low-temperature thermoluminescence peak which is usually attributed to the unfreezing of motion in the polytetrafluoroethylene amorphous phase, since its intensity changes inversely with the degree of crystallinity. According to Nikolskii and Buben [127], it is located at

148 K, whereas Tomita [122] observed it at 206 and 177 K for samples irradiated in vacuum and air, respectively. The most surprizing observation, however, was that made by Mele *et al.* [121], who found that the low-temperature peak between 146 and 177 K appears only in the presence of dissolved gases such as oxygen or helium. According to Partridge [4], this suggests the existence of a "molecular cavity" type of electron trap formed by a molecule, or molecular group, that does not have a positive electron affinity but that can nevertheless trap electrons when it is in a polymer matrix, perhaps by distorting the polymer structure in its immediate vicinity to form a sort of cavity. Since helium could not by itself trap positive or negative charges, trapping must be a cooperative effort between helium atoms, polymer chains, and electrons. This certainly interesting phenomenon requires more detailed examination. A larger number of gases should be studied in an attempt to establish more quantitatively the relationship between thermoluminescence and gas adsorption.

6.4   Multicomponent Polymer Systems

6.4.1   Compatibility of Polymers

Over the years, there has been a great deal of interest in studies of the structure and properties of multicomponent polymer systems. These materials are formed by combining two or more polymers by various methods, such as mechanical blending, solution casting, or direct chemical synthesis. The morphology of the resulting products depends to a large measure on the compatibility of the constituents. *Chemical compatibility* is defined in classical chemistry as complete molecular mixing. With respect to macromolecules, whose molecular size can be large, this definition loses its significance, since many heterogeneous systems would be considered homogeneous even though separate phases can be proven to be present. Thus, for polymers, mixing of segments rather than polymer chains should be of concern.

Flory [128] was probably the first to clearly state that two polymers are mutually compatible with one another only if their free energy of interaction (ΔG_m) is negative; that is,

$$\Delta G_m = (\Delta H_m - T\Delta S_m) < 0 \tag{6.6}$$

where ΔH_m and ΔS_m represent the enthalpy and entropy of mixing, respectively. The enthalpy of mixing depends on the strength of unlike 1–2 contacts in the mixture relative to the like 1–1 and 2–2 contacts in the pure components. In the usual case of dispersion forces, the heat of mixing of organic molecules is generally positive, so that the difference in energy between a pair of like and unlike polymer segments is negative. For nonpolar molecular species, the ΔH_m term in Eq. (6.6) is positive, opposing mixing. This is true for large and small molecules alike. While the ΔH_m values for mixing monomer and polymer species obviously differ, the changes tend to be modest. However, there is a dramatic difference in entropy contributions between small and large molecules. It is obvious that a mixture is more "disordered" than the pure components. Thus the entropy contribution ΔS_m is positive, and this is the main reason for the solubility of small-molecule liquids. However, entropy depends on the number of molecules in the system. When high-molecular-weight polymers are mixed, the combinatorial entropy becomes relatively unimportant and the main cause of solubility for small molecules is no longer available. According to theory, for high molecular weights, compatibility can be achieved only through a favorable contribution from specific interactions, e.g., ion–dipole interactions, hydrogen bonds, or a charge-transfer complex.

It must be realised that for polymers to be compatible, it is necessary, but not sufficient, to satisfy the condition $\Delta G_m < 0$, a fact that is not always sufficiently stressed. Upon somewhat closer scrutiny of the situation, it appears that with many partially compatible systems, ΔG_m may well be negative over the entire composition range, phase relationships then being governed by subtle details of the composition dependence of ΔG_m [129]. The deciding factor is then the concentration dependence of ΔG_m, and a compatible binary polymer system must satisfy the stability criterion

$$[\partial^2 \Delta G_m / \partial v_1^2]_{T,p} > 0$$

where v_i is the volume fraction of one of the components.

The thermodynamic approach involves the application of equilibrium conditions to systems that, due to their high viscosity, may not be able to reach equilibrium within the time of experiment. As a consequence, some metastable compositions could remain in their state almost indefinitely if the polymer mixtures are very viscous. It was suggested that this is one possible reason that some of these compositions sometimes appear as a single phase and at other times show phase separation [130].

The most important question from both theoretical and practical points of view is determination of the character of the interface between different phases. Even with small molecules (oil/water emulsion, for instance), the composition boundary does not consist of a vertical composition cliff, with diffusion and entropic forces causing at least a few angstroms of interphase width. Since monomers are always compatible to a much greater extent than corresponding polymers, solubility on the segmental level within a phase boundary of a finite width can be readily assumed for insoluble polymer–polymer systems. In the vicinity of a boundary surface the chains cannot be oriented freely. It is necessary to consider the entropy loss that accompanies this sort of restriction of chains. Basically, the equilibrium of the boundary surface depends on a balancing of the contact energy and the restricting entropy loss. When some degree of interpenetration occurs in an interfacial region between polymers A and B, it produces an energy-of-mixing contribution to the interfacial tension. The *interfacial tension* is the free energy per unit surface area, and it indicates the extent of interdiffusion of the different molecules in the boundary surface. It has been calculated that this value is on the order of a few ergs per square centimeter and the thickness of the boundary surface is in the region of a few tens of angstroms [131]. The asymptotic phases A and B in contact remain essentially pure. In the interface, however, mixing occurs as the concentration of one polymer falls, being replaced by the other. Since the actual phase dimensions can be small, the interface volume can be appreciable, with important consequences for the mechanical behavior of the material. The phase interaction at the boundary governs the adhesion between the domains of the phase-separated polymeric composites and thus is responsible for stress transfer across the interface, which is needed for the attainment of physical strength.

It should be emphasized that all the above-mentioned considerations of compatibility relate to amorphous polymers. When one or both components of a blend are crystalline, the preceding considerations apply only to their amorphous phase. The compatibility of polymers in their crystalline phase, i.e., the formation of a co-crystal, is very unlikely and has not yet been conclusively proven [132].

6.4.2 Experimental Methods for Evaluating Compatibility

Despite recent advances, the bulk compatibility of polymers cannot be predicted by theory. Overall, it can be said that in view of the approximate character of the theories, and particularly in view of the difficulties in obtaining accurate experimental values for the required parameters, only guideline qualitative (but not precise quantitative) predictions can be made for the compatibility of polymers.

The study of compatibility is further complicated by the fact that so far no direct experimental method is available for the measurement of polymer compatibility in the melt or the solid state. Cloud-point measurement and light-scattering analysis can be used to determine the position of the binodal [133] or spinodal [134]. However, a crossing, or at least a close approach, of the transition from the homogeneous to the heterogeneous state, or vice versa, is required in these measurements. This is possible only if the transition temperature is below the temperature at which decomposition of the component polymers occurs. Further restrictions are imposed by the size of macromolecular domains and the refractive indexes of the components. A system with domains smaller than the wavelength of visible light cannot be evaluated by these methods. Also, if both polymers have the same refractive index, the system would always be transparent independent of the degree of heterogeneity.

Other techniques also have essential limitations. The validity of measurements made on polymer solutions (mutual-solvent method) rests on the assumption that the solubility gap extends far into the polymer 1–polymer 2–solvent solubility triangle. Systems exhibiting closed solubility gaps do not meet this criterion, and a great number of polymer blends cannot be analyzed by this technique [135].

Electron microscopy and small-angle X-ray scattering can yield some information on the structure and morphology of a multiphase system, but they do not provide any direct data on the compatibility of two polymers. When viewed on a small enough scale, even homopolymers are anisotropic and hence for some purposes can be considered as heterogeneous systems. Heterogeneities are also observed in polymer mixtures, but this is not an indication of mutual insolubility, since a complete structural uniformity is not a necessary condition for compatibility [136].

The most widely used method for evaluating compatibility is determination of the transition temperatures. Compatible polymers are considered to be those which form a single-phase blend with a single glass-transition temperature dependent on the composition of the blend. Polymer blends which exhibit two or more T_g values corresponding to the glass-transition temperatures of the individual components are considered to be incompatible. In a two-component incompatible system, two distinct T_g

values are usually observed when the phase domains are larger than several hundreds of angstroms. For smaller domains, the interfacial contribution to relaxation becomes significant. The glass-transition behavior changes from two distinct transitions to one broad, intermediate transition in a systematic manner. A broadened transition can result even in thermodynamically compatible mixtures if the minimum number of mer units required for independent contribution to the relaxation spectrum is subject to wide composition variations. At least 50 mers must undergo coordinated motion for a glass-transition relaxation to occur [137]. A volume of approximately $10.000A^3$ is involved and thermal fluctuations are significant when major relaxing molecular rearrangements are localized in regions of such small volume. Only volumes of $100.000A^3$ or more have "average" composition.

Evaluation of compatibility on the basis of the transition-temperature criterion has been successfully used when optical and other methods give doubtful results. In a number of cases, however, what is described as compatible by one method is not by another. For example, a blend of poly(2,6-dimethyl-1-1,4-phenylene ether) with atactic polystyrene has two relaxation peaks corresponding to the constituting polymers in dynamic-mechanical measurements, but it has a single T_g when it is determined by differential thermal analysis [138]. These situations are relatively rare and may occur in multicomponent systems with small phase dimensions. This indicates that different methods respond to relaxations of different molecular subsystems and that this becomes more apparent when phase domains are not much larger than the critical domain size necessary for cooperative relaxation. T_g measurements determined by mechanical spectroscopy or any other means of applying stress are based on the fact that the motions under stress are strongly dependent on the free volume. However, observing a T_g value by a method based on a change in temperature (no stress) results in a different molecular superstructural reorganization than would a T_g value observed isothermally by a stress-relaxation experiment. Thus is seems that static and dynamic experiments represent different types of molecular motion [139]. If this is the case, the equivalent-frequency principle cannot be fundamentally correct and the extrapolation of T_g values obtained by high-frequency dynamic methods to lower frequencies characteristic for static measurements requires some reservation.

6.4.3 Polymer Blends

6.4.3.1 Homogeneous and Heterogeneous Polymer Blends

The transition-temperature criterion for compatibility evaluation provides the ground for application of the radiothermoluminescence technique to the study of polymer blends. In the heterogeneous blend comprising two separate phases, the fate of a secondary electron generated on radiation exposure will depend on the size and shape of the domains and on the average travel range of secondary electrons [140]. This can be rationalized by using Fig. 6.21, a schematic representation of a region near the interface between domains of the phase-separated blend components polymers A and B. When both polymer phases are in the glassy state, the secondary electron generated in and traveling through the frozen matrix A will most likely be trapped at T_A in phase A. If the secondary electron is generated near the AB phase boundary,

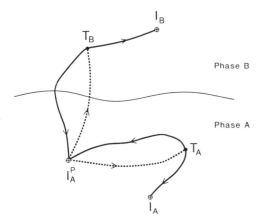

Fig. 6.21. Ion recombination process. Schematic drawing of electron trapping near the domain interface in a heterogeneous polymer blend. After [140], © 1979, American Chemical Society.

it can, however, cross the interface and be trapped at T_B in phase B. One can make a similar argument for the trapping of secondary electrons generated in phase B.

On subsequent release of the secondary electrons from the traps on warm-up, the electrons trapped in phase A should more or less all combine with ions present in the same phase. It will very likely be parent ion I_A^P from which the electron originated or another ion I_A in the vicinity of the trapped electron. Those electrons which transferred and were trapped in phase B could in principle recombine with ions I_B of phase B. Again, however, it is probable that most of these electrons, attracted by the Coulomb field of their parent ions, will return to phase A and recombine there. The reason for this is that once an electron is released from one of the traps, its energy is only kT. Without much of its own initiative, its travel and range are governed by the Coulomb field of ions in its vicinity. The separation between positive and negative ions required to reduce the energy of Coulomb attraction to thermal energy kT at $T = 77$ K is about 1000 Å. Since the average distance of initial charge separation is estimated to be 60 Å [141] and the average distance between two ions of about 200 Å [135], we can conclude that the parent ion, on average, will be the closest neighbor to the trapped electron. Hence it is likely to recombine with it. Thus it can be assumed that at T greater than T_g^A and T_g^B, essentially all the electrons created in phase A will have returned to A and most electrons created in phase B will have returned to B.

For domains much larger than 60 Å—the average travel distance of the secondary electron following irradiation—practically all electrons generated in each phase will be trapped in this phase and the glow curve of the blend can be expected to be the sum of the glow curves A and B. Yet the glow curve of the blend may not be the sum of A and B. This is so because the electrons crossing the phase boundary from B to A will do so at a transition temperature characteristic of polymer B. However, this effect may be pronounced only when phase domains are not essentially larger than 60 Å. Thus a deviation of the glow curve of the blend from that of A plus B can be visualized even in the hypothetical case of small phases divided by geometric bourders. In a real system with diffusive interfaces between domains, the appearence of such diviation will depend not on the phase dimensions, but rather on the volume contribution of the interface.

Bohm and Lucas [140] showed that in a polymer blend the existence of the interface can be substantiated and its volume contribution measured by means of radio-thermoluminescence. They expressed the glow curve of a heterogeneous system with the interface as the sum of the contributions from regions pure in A and B as well as from the interface

$$I(T) = X_A I_A(T) + X_B I_B(T) + \sum_1^n X_m I_m(T) \tag{6.7}$$

where X_A and X_B are the volume fractions, and I_A and I_B are the luminescence intensities of pure A and B in the blend. For the purpose of estimating the glow curve of the interface, it was thought of as an ensemble of n regions of composition X_m. Regions of equal composition may exist in a heterogeneous polymer system at different locations. However, regardless of where they are relative to each other, they should exhibit the same type of glow curve. Hence the joint contribution of all these regions to overall luminescence intensity is $X_m I_m(T)$, where X_m is the volume fraction of material in all these regions, and $I_m(T)$ is the normalized luminescence intensity of a homogeneous blend with composition X_m. The total contribution of the interface is then the sum of the contributions from the entire ensemble of regions, as expressed in Eq. (6.7). The glass-transition temperature of the interface regions of different composition was assumed to follow the relation

$$T_g^{AB} = X_A T_g^A + X_B T_g^B \tag{6.8}$$

And the intensity of light given off on passing through the T_g maximum is assumed to be the sum over the weighted intensity contributions of the pure blend components measured at their respective T_g values. Then the total contribution of the interface can be calculated by

$$I^{IF}(T) = \sum_1^n X_m \left[X_m^A I^A(T') + X_m^B I^B(T') \right] \tag{6.9}$$

following a transposition of the glow curves $I(T)$ for the polymers A and B to $I(T')$ by a transformation of the temperature coordinate according to

$$T' = T + (T - T_1)k^{AB}$$
$$k^{AB} = (T_g^m - T_g^{AB})/(T_g^{AB} - T_1) \tag{6.10}$$

where T_1 is the irradiation temperature, and T_g^m is the equivalent of T_g^{AB} after transformation of the temperature coordinate T to T'. Since the total number of photons emitted is not altered by transposition of the glow curves, that is,

$$\int_{T_1}^{T > T_g^{AB}} I(T)\, dT = \int_{T_1}^{T > T_g^{AB}} I(T')\, dT'$$

and the intensity-correction function, $f(T - T_1) = I(T')/I(T)$ can be obtained.

The material balance in the pure phases and in the interface gives

$$X_A^0 + X_B^0 = 1$$

$$X_A = X_A^0 - X_A^{IF} \qquad X_B = X_B^0 - X_B^{IF} \tag{6.11}$$

$$X^{IF} = \sum_1^n X_m = X_A^{IF} + X_B^{IF}$$

where X_A^0, X_A, and X_A^{IF} are the volume fractions of A in the entire sample, in the regions pure in A, and in the interface, respectively, and X^{IF} and X_m are the volume fractions of the interface and of the mth part of the ensemble of interface regions having a composition X_m^A and X_m^B.

Some of the assumptions made by Bohm and Lucas require reservations, in particular Eq. (6.8). Nevertheless, using Eqs. (6.7)–(6.11), X^{IF} and X_m^A values can be estimated from known values of X_A^0 and experimental glow curves for the blend and the component polymers comprising it.

When a combination of noncompatible polymers is mixed, the morphology of the blend depends on the ratio and relative viscosities of the components. The component that is present in small amounts forms a dispersed phase. When both the components are mixed in more or less equal proportions, the component which has the lower viscosity tends to form a continuous phase more readily. When equal amounts of components with about the same viscosities are mixed, both phases may become continuous. Usually, the size of polymer domains is in the region of microns and the width of the interface does not exceed several tens of angstroms. Thus the interface occupies fractions of a percent of the total blend volume, and for many heterogeneous polymer systems, the glow curve can be well approximated by the sum of the glow curves of the components [11,136].

In particular, the glow curves of heterogeneous two-component blends possess two thermoluminescence maxima located in the glass-transition temperature regions of the individual components. The variations in the ratio of the components result in a gradual increase in intensity of one of the maxima with a corresponding weakening of the other. This is illustrated in Fig. 6.22, which presents the glow curves for *cis*-polybutadiene–ethylene-propylene copolymer blends of different compositions [142]. The T_g values determined from the positions of the maxima were 175 and 209 K over the whole composition range and coincided with those of *cis*-polybutadiene and ethylene-propylene copolymer, respectively. Moreover, the width of the two main peaks were about equal to those of the pure materials, hence indicating that no appreciable interface exists between the domains of the component polymers. The heterogeneous nature of the polybutadiene–ethylene-propylene copolymer blend was confirmed by a transmission electron microscope examination [140]. A photomicrograph of the blend comprising equal parts of the components shows discrete, sharply outlined domains of ethylene-propylene copolymer approximately 1.5 μm in size embedded in a continuous matrix of polybutadiene. The polybutadiene–ethylene-propylene copolymer blend is a typical example of a heterogeneous system with practically no phase interaction. Thus the intensity of luminescence originating in separate phases is proportional to their volume fractions. This provides the linear dependence of the intensity of each of the T_g maxima on the concentration of the components.

In some cases, however, a deviation from linearity has been noted [143]. Evalu-

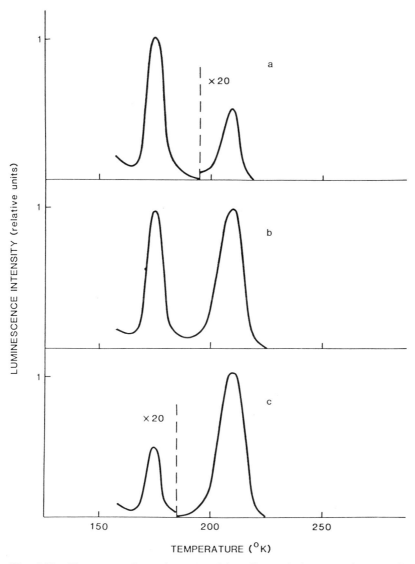

Fig. 6.22. Glow curves for various *cis*-polybutadiene–ethylene-propylene copolymer blends. The weight fractions of *cis*-polybutadiene in the blends are *(a)* 0.97, *(b)* 0.50, and *(c)* 0.03, respectively. After [142], with permission of the Rubber Division of the American Chemical Society.

ation of a polybutadiene–polyisoprene blend by means of radiothermoluminescence confirmed its heterogeneous nature, which followed from the fact that two T_g values (190 and 220 K) coinciding with the corresponding values for the individual components were observed in the entire range of concentrations (1–99%). Despite the obvious heterogeneity, the proportionality between the intensities of the thermoluminescence peaks at 190 and 220 K and the concentration of the components was altered: The intensity of the peak at 220 K was lower than the value expected from the linearity, whereas the peak at 190 K showed the opposite tendency.

There are several factors which may cause this deviation from the linearity, even in a truly heterogeneous system. First, as was mentioned earlier, such an effect may be expected in a system with very small domains where some of the charges originating in one phase undergo trapping and subsequent detrapping in the other phase. Second, this effect can be due to the detrapping of electrons located in the phase with the higher T_g value as a result of the light emitted by the other phase undergoing the glass transition. Third, this effect can be caused by the presence of an interface region between the separate phases. The latter possibility was stated by Melnikova et al. [143] as the reason for the deviation from linearity in the transition intensity–composition plot. However, this can be disregarded because the developed interface is expected to bring about changes not only in the relative intensities, but also in the positions and shape of the individual maxima. Also, the interface concept does not provide an explanation for the increase in intensity of the transition at 190 K. If the major cause for the variations in luminescence intensity at 190 and 220 K was the stabilization in one phase of some of the charges originating in the other phase, nonobvious preferential crossing from the polyisoprene to the polybutadiene phase has to be assumed. However, bleaching of some of the electrons stabilized in the polyisoprene phase does account for the observed effect and thus can be considered as the most probable cause of the redistribution of the thermoluminescence intensities at 190 and 220 K.

Several radiothermoluminescence studies have been performed on the compatibility of polybutadiene with butadiene-styrene copolymer [11,14,135,140,144]. This blend provides a rare system in which phase uniformity can be varied by changing the conditions of preparation and heat treatment. Figure 6.23 shows the glow curves of polybutadiene, butadiene-styrene copolymer (styrene/butadiene ratio 20/80), and a blend comprising 50 wt% of each of the elastomers prior to and after annealing at 423 K [135]. As can be seen, the glow curve of the blend even prior to annealing cannot be approximated by the sum of the glow curves of the components. The essential broadening and overlapping of the transition peaks indicate a certain degree of solubility of the components even at room temperature. After annealing, instead of the two dominant transitions at 178 and 202 K, only one strong peak at 191 K can be noted and, on its shoulder, a less pronounced transition at about 197 K. Thus the compatibility of the polymers is considerable enhanced by annealing, which leads to the formation of a pseudohomogeneous system possessing an ensemble of microregions which differ in composition. This is characteristic for a system with a developed interface.

The thermodynamic calculations performed by Scott [145] showed that mixtures of copolymers form a single-phase system if the difference in their solubility parameters $\Delta\delta \leqslant 0.055$ ($T = 300$ K, $\rho = \mathrm{lg\ cm}^{-3}$, Mw = 100.000). For butadiene-styrene copolymers, this condition is satisfied if the variation in their composition is not greater than 7% [146]. Thus the increase in styrene content in the butadiene-styrene copolymer is expected to diminish its compatibility with polybutadiene. Indeed, the butadiene-styrene copolymer with 30% styrene is compatible with polybutadiene at room temperature to a lesser extent than the copolymer with 20% of styrene. Nevertheless,

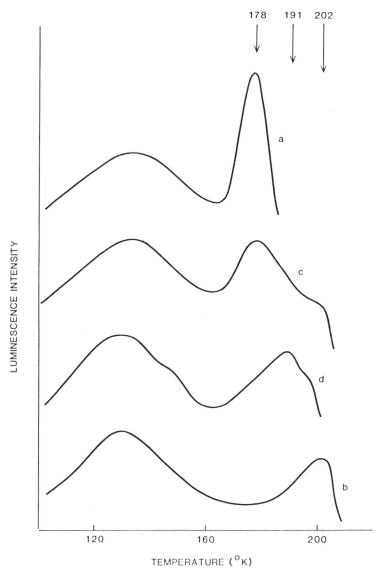

Fig. 6.23. Glow curves for *(a)* polybutadiene (cis/trans/vinyl = 54/35/11) and *(b)* poly(butadiene-co-styrene) (20% styrene) and a blend comprising 50 wt% of each of the two elastomeres prior to *(c)* and after *(d)* annealing in a vacuum at 423 K for 26 hours. After [135], with permission of the Rubber Division of the American Chemical Society.

as in the case described above, its compatibility with polybutadiene can be enhanced by annealing at 423 K [144]. It has to be noted that the homogenization achieved on annealing decreases during subsequent room-temperature aging (Table 6.8). On prolonged aging, the blend tends to revert to the initial state at room temperature, but over a far longer time than that required for reaching the equilibrium at 423 K (which

Table 6.8. Positions of low- and high-temperature radiothermoluminescence maxima for mixtures of *cis*-polybutadiene and butadiene-styrene copolymer (30% styrene) heated for 15 min at 150°C in relation to subsequent holding time at room temperature

Holding time (hours)	Position of low-temperature maximum (K)	Position of high-temperature maximum (K)
0	184	—
0.03	182	203
0.17	181	205
1	181	206
3.5	181	209
20	180.5	211
110	180	212.5
336	179.5	215.5
Before heating	179	218

Source: Ref. 144, Plenum Publishing Corporation, with permission.

was estimated to be at least 40 min [144]). A considerably increased solubility of the two polymers at the annealing temperature is most probably caused by a reduction in the free energy of mixing on increase in temperature, as might be expected from operation in an upper critical-solution temperature regime. Annealing in a vacuum of polybutadiene–butadiene-styrene copolymer blends of several weight ratios at various temperatures and subsequent evaluation by means of radiothermoluminescence provided Sershnev and Pestov [147] with data sufficient for obtaining the phase diagram of this system (Fig. 6.24). All points located above the curve correspond to formation of a single-phase blend. The critical composition and upper critical-solution temperature were found to be equal to 37 wt% and 423 K, respectively.

Finally, let us present some of the results obtained for the blend of *cis*-polybutadiene with high-vinyl polybutadiene [13,136]. In this case, the glow curves of blends of different compositions are characterized by a single maximum situated between the maxima of the individual polymers (Fig. 6.25). Similar glow curves were obtained independently of the method of preparation (roll-mill mixing or solvent evaporation). Neither quenching of the blends to liquid nitrogen temperature nor prolonged annealing at 423 K changed the shape of the curves. In accordance with the glass-transition-temperature criterion of compatibility, it can be concluded that *cis*-polybutadiene and high-vinyl polybutadiene are compatible in a whole range of concentrations.

The evaluation of a variety of polymer blends showed that separate, mutual solubility of two polymers with a third does not predetermine their mutual solubility in each other [136].

6.4.3.2 Dependence of the Glass-Transition Temperature on the Homogeneous System Composition

There are several equations for calculating the glass-transition temperatures of copolymers and homogeneous polymer blends. Among them, the most widely used are

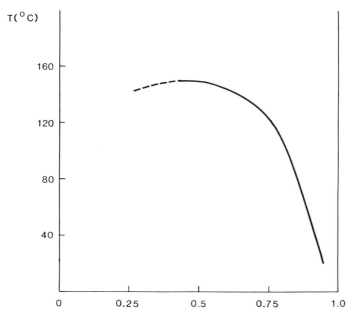

WT. FRACTION OF BUTADIENE-STYRENE COPOLYMER

Fig. 6.24. Phase diagram for a polybutadiene–butadiene-styrene copolymer (30 wt% styrene) blend as revealed from the radiothermoluminescence data [147].

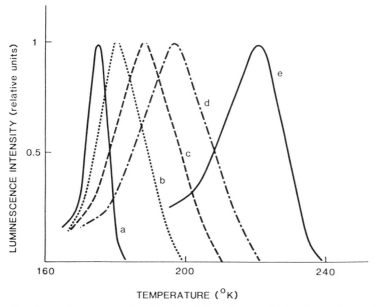

Fig. 6.25. Glow curves for *(a)* *cis*-polybutadiene and *(e)* polybutadiene with 60% vinyl content and their blends with *cis*-polybutadiene weight fractions of *(b)* 0.7, *(c)* 0.5, and *(d)* 0.3. After [13], Pergamon Journals, Inc., with permission.

those of Gordon and Taylor [148] and Fox [149]. Figure 6.26 shows the glass-transition temperature as a function of composition for a *cis*-polybutadiene–high-vinyl polybutadiene blend as obtained by the radiothermoluminescence method. Curve *A* represents the Gordon-Taylor equations:

$$T_{gm} = \frac{W_1 T_{g1}(\alpha_{L1} - \alpha_{G1}) + W_2 T_{g2}(\alpha_{L2} - \alpha_{G2})}{W_1(\alpha_{L1} - \alpha_{G1}) + W_2(\alpha_{L2} - \alpha_{G2})} \tag{6.12}$$

which was derived for copolymers rather than polymer blends. In this equation, T_{gm} is the glass-transition temperature of the copolymer (the glass-transition temperature of the blend in our case), T_{g1} and T_{g2} are the glass-transition temperatures of *cis*-polybutadiene and high-vinyl polybutadiene, W_1 and W_2 are the weight fractions of

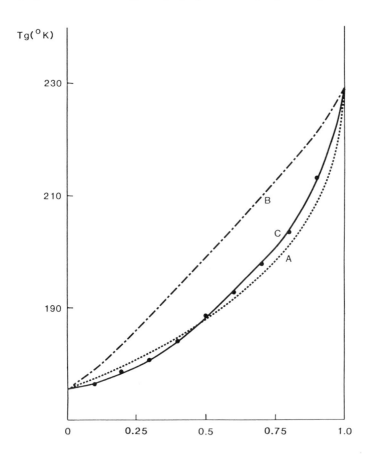

WT. FRACTION OF HIGH–VINYL POLYBUTADIENE

Fig. 6.26. The glass-transition temperature as a function of composition for a *cis*-polybutadiene–high-vinyl polybutadiene blend: *(A)* Gordon-Taylor equation; *(B)* Fox equation; and *(C)* Equation (6.14). After [142], with permission of the Rubber Division of the American Chemical Society.

the corresponding elastomers in the blend, and α_{L1}, α_{G1}, α_{L2}, and α_{G2} are the thermal-expansion coefficients of polymers in the following order: *cis*-polybutadiene above the glass-transition temperature, *cis*-polybutadiene below the glass-transition temperature, high-vinyl polybutadiene above the glass-transition temperature, and high-vinyl polybutadiene below the glass-transition temperature. The dilatometric measurements were carried out over the temperature range from 100–270 K [142]. It was found that $\alpha_{L1} = 1.0 \times 10^{-4}$, $\alpha_{G1} = 0.53 \times 10^{-4}$, $\alpha_{L2} = 0.63 \times 10^{-4}$, and $\alpha_{G2} = 0.5 \times 10^{-4}$ cm^3/g·deg. Curve *B* represents the Fox equation, which was originally derived for both copolymers and mixtures of polymers with plasticizers:

$$T_{gm} = T_{g1}T_{g2}/(W_1T_{g1} + W_2T_{g2}) \tag{6.13}$$

Figure 6.26 clearly shows that the data do not fit the Fox equation. However, the Gordon-Taylor equation, which takes into account not only the glass-transition temperatures of the components, but also the differences in their thermal-expansion coefficients, gives a better corelation with the experimental results.

An even better fit is obtained when the empirical relation

$$T_{gm} = T_{g2} - (T_{g2} - T_{g1})W_1^c \tag{6.14}$$

is utilized [150]. In Eq. (6.14), W_1 is the weight fraction of the component with the lower glass-transition temperature, and c is a constant for a particular polymer blend. In Fig. 6.26., curve *C* represents Eq. (6.14), with the constant c equal to 0.44. The successful application of the Eq. (6.14) to the evaluation of glass-transition temperatures in other homogeneous polymer blends was demonstrated elsewhere [150].

6.4.3.3 Effect of Cross-Linking One of the Components on Compatibility

Since the formation of a homogeneous blend requires diffusive interpenetration of the components' macromolecular chains, the following question can be asked: What degree of cross-linking would prevent homogenization and to what extent? In order to get such information, *cis*-polybutadiene was cross-linked by γ-radiation and then mixed with high-vinyl polybutadiene at a 30:70 wt% ratio [151]. Mixing was performed on roll mills for 15 min. It was found that independent of the degree of cross-linking of *cis*-polybutadiene, further mixing did not cause any changes in the glow curves. This shows that even if destruction processes took place during milling, they did not substantially affect the degree of dispersion of the blend.

Glow curves for blends with different degrees of cross-linking of *cis*-polybutadiene are given in Fig. 6.27. The arrows on the abscissa show the T_g values of the individual components. It is seen from the curves that a single-phase homogeneous system obtained by mixing of un-cross-linked or slightly cross-linked polymers turns into a two-phase (or multiple-phase), partially compatible system when a certain degree of cross-linking of *cis*-polybutadiene is attained. As the degree of cross-linking increases, the heterogeneity of the blend becomes more evident. Such blends include two phases with higher and lower T_g values. The composition of the high-T_g phase depends on the degree of cross-linking of *cis*-polybutadiene and may be calculated

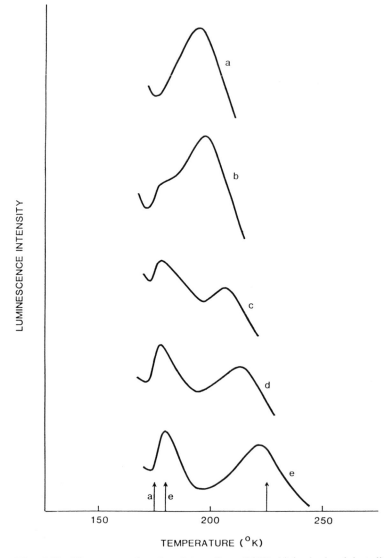

Fig. 6.27. Glow curves for *cis*-polybutadiene (30%)–high-vinyl polybutadiene (70%) blends. The number of linked chains per cubic centimeter (N_c) in *cis*-polybutadiene: *(a)* less than 0.8×10^{19}; *(b)* 1.24×10^{19}; *(c)* 1.68×10^{19}; *(d)* 2.2×10^{19}; and *(e)* 10.1×10^{19}. After [151], Plenum Publishing Corporation, with permission.

using the compositional dependence of T_g for the blend of un-cross-linked components. The second, low-T_g phase of the blend consists almost completely of cross-linked *cis*-polybutadiene.

The fact that cross-linked *cis*-polybutadiene plasticizes high-vinyl polybutadiene indicates that the network structure of cross-linked rubber is not the simple homogeneous composition generally assumed. The part of *cis*-polybutadiene with the higher concentration of cross-linkes forms an individual phase. The other part of it, with the relatively low concentration of cross-links and, correspondingly, with a larger amount of molecular chains attached to cross-links at one end only and, consequently, able to diffuse (inactive chains, according to Flory [152]), plasticizes high-vinyl polybutadiene. In addition, the sol fraction of cross-linked *cis*-polybutadiene plasticizes high-vinyl polybutadiene as well.

In order to obtain quantitative information on the amount of cross-linked *cis*-polybutadiene able to form a homogeneous blend with high-vinyl polybutadiene, the *coefficient of interdiffusion* was introduced. The latter is defined as the ratio of the cross-linked component blended homogeneously with the un-cross-linked component to the actual amount of cross-linked component used. Thus, for complete mutual solubility of the components, the coefficient of interdiffusion was taken as 1.

Figure 6.28 shows how the coefficient of interdiffusion of the system depends on the number of linked chains per unit volume (N_c) in *cis*-polybutadiene. This relationship is plotted for systems containing both sol and gel fractions of *cis*-polybutadiene (curve 1) and is also recalculated to take into account only the gel fraction (curve 2). For a low concentration of cross-links ($0 < N_c \times 10^{-19} < 0.8$), both curves have a common linear portion in which the coefficient of interdiffusion is 1. As the concentration

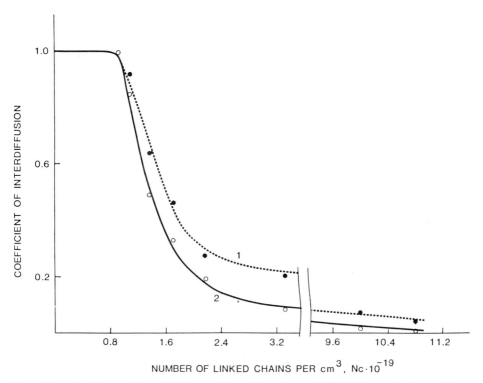

Fig. 6.28. Coefficient of interdiffusion of *cis*-polybutadiene (30%)–high-vinyl polybutadiene (70%) blends as a function of the number of linked chains per cubic centimeter *(N_c)* in *cis*-polybutadiene: (1) *cis*-polybutadiene containing sol and gel fractions; and (2) *cis*-polybutadiene containing only a gel fraction. After [151], Plenum Publishing Corporation, with permission.

of cross-links increases above this level, the coefficient of interdiffusion begins to fall—rapidly at first ($0.8 < N_c \times 10^{-19} < 2.2$) and then more slowly ($N_c \times 10^{-19} > 2.2$). The existence of the initial linear portion of the curve indicates that up to a certain density of cross-linking, the coefficient of interdiffusion is still 1. This can be explained by assuming either that the network is not continuous or that *cis*-polybutadiene with infrequent cross-links is capable of swelling and absorbing the second component (which in our case behaves as a high-molecular-weight solvent) in an amount equaling or exceeding that corresponding to the initial ratio.

Thus blends with different degrees of phase in-homogeneity can be obtained by changing the degree of cross-linking of one of the components prior to mixing.

6.4.3.4 Other Factors Influencing Compatibility

Most polymers are incompatible. Therefore, the task of decreasing compatibility by cross-linking one of the components is rather of academic interest. However, the opposite goal—namely, the development of methods capable of providing various levels of compatibility in blends of originally incompatible polymers—is of significant practical importance because it is well established that optimum mechanical properties are achieved at a certain degree of separation into distinct phases, i.e., at a certain degree of homogeneity.

There are several reports indicating that various levels of homogeneity can be obtained when a heterogeneous polymer blend is subjected to shear deformation under high pressure [153–155]. The combined effect of high pressure and shear is achieved by using a Bridgeman anvil-type apparatus, in which the rotation of the upper anvil provides plastic shearing proportional to the angle of rotation. It was shown that deformation at pressures higher than 1 kbar modifies the phase structure of the blends. The convergence of the glass-transition temperatures of the individual components is already observed at anvil-rotation angles as small as $100–200^0$ (Fig. 6.29). Larger deformations lead to transformation to a homogeneous system characterized by a single glass-transition temperature value intermediate to those of the components. It is interesting that the homogeneous sample (Fig. 6.29, curve *d*) exhibits the half-width of the main thermoluminescence peak of 17°, i.e., less than the half-widths of the corresponding peaks in polyethylene and polypropylene. Similar results were obtained for polypropylene–polyethylene, polypropylene–ethylene-propylene copolymer, and polypropylene–polyisobutylene blends, although the conditions for a complete homogenization varied from one system to another. Homogeneous samples obtained at pressures less than 10 kbar returned to the heterogeneous state after annealing at 220 C, whereas samples deformed at pressures greater than 10 kbar retained the single-phase structure.

Shear deformation under high pressure may influence various changes in polymeric blends, such as disruption and reconstruction of the supermolecular formations, destruction and orientation of macromolecules, and intermolecular cross-linking. Although each of these factors may affect relaxation, none of them taken separately can precipitate the formation of a single-phase system. Most certainly their superposition is the driving force of homogenization. This can be visualized as the process during

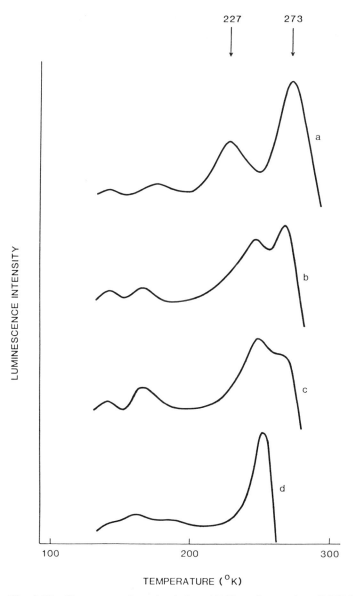

Fig. 6.29. Glow curves for polyethylene (10%)–polypropylene (90%) blends: *(a)* initial blend; and *(b–d)* blends exposed to shear deformation under 20-kbar pressure and anvil-rotation angles of *(b)* 500, *(c)* 1000, and *(d)* 1500°. The arrows indicate the transition temperatures of individual components. After [154], Pergamon Journals, Inc., with permission.

which individual phases lose their ''identity'' due to a very fine level of dispersion (not exceeding 8–10 nm), and the intermediate structure is then fixed by very high viscosity and/or intermolecular cross-linking.

Although the large-scale realization of homogenization by shearing under high pressure may not be visable, this process can, in principle, be used for the production of compatibilizing agents for blends of respective polymers. It seems to be equivalent to the formation of a compatibilizing agent in suti when incompatible polymers are subjected to shearing forces which rupture polymer chains to generate radicals which undergo coupling [156].

6.4.3.5 Covulcanization of Elastomers

The history of polymer blends in the rubber industry is long. Elastomer blends are compounded with carbon black, softening agents, vulcanizing agents, accelerators, etc., and so as systems they are very complex indeed. Hence a formulator faces a great many problems of common and specific nature. In particular, the former includes evaluation of the compatibility of the components, whereas the latter deals with the distribution of vulcanizing agents and accelerators, which, in some cases, may cause an imbalance in the relative rates of cross-linking in different phases.

The evaluation of a variety of elastomer blends by means of radiothermoluminescence showed that vulcanization does not influence any change in phase uniformity: The homogeneity or heterogeneity possessed by a blend before vulcanization is retained afterwards [11,136].

Mixing of rubbers with vulcanizing agents sharply decreases the intensity of thermoluminescence. The lower luminescence output is first determined by the fact that mixing with constituents makes the blend nontransparent. The second, more significant, cause is that vulcanizing ingredients effectively extinguish luminescence. The addition of sulfur or substances containing the extinguishing groups OH, Br, or Cl sharply reduces light output during the recombination of certain active centers. Vulcanization somewhat increases the intensity of the thermoluminescence of the vulcanizate compared to an unvulcanized mixture most probably because of the lower extinguishing ability of the ingredients in combined form. Although the addition of vulcanizing ingredients to elastomers leads to a significant decrease in luminescence intensity, it scarcely changes the temperatures and half-widths of the thermoluminescence maxima [136].

A comparison of glow curves before and after vulcanization allows one to determine the shift in T_g values due to cross-linking. Using these data, the concentration of cross-links can be calculated and compared with the corresponding results for individual components, thus providing information about the uniformity of distribution of cross-linking agents.

Although cross-linking by itself does not change the level of homogeneity of a blend, it fixes the structure reached by the system at the temperature of vulcanization. The homogeneity of a multicomponent polymer system exhibiting an upper critical-solution temperature increases with an increase in temperature. Then, provided that the rate of homogenization is higher than the rate of cross-linking, a larger degree of phase uniformity is achieved by the system at elevated temperatures. The formation of cross-links between macromolecules restricts their ability to move freely and prevents phase separation on subsequent cooling to room temperature.

As was mentioned earlier, a *cis*-polybutadiene–styrene-butadiene copolymer blend, although heterogeneous at room temperature, exhibits an upper critical-solution temperature at 150°C. Therefore, vulcanization of this blend at elevated temperatures could be expected to promote homogenization. Different vulcanization processes have been studied in this respect by Nikolskii *et al.* [14]. Glow curves for *cis*-polybutadiene–butadiene-styrene copolymer blends cross-linked with sulfur are presented in Fig. 6.30. For comparison, Fig. 6.30 shows the glow curve of the crude blend. It

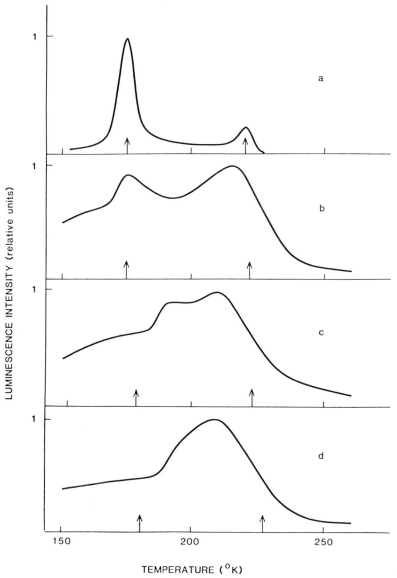

Fig. 6.30. Glow curves for an un-cross-linked blend of *cis*-polybutadiene with butadiene-styrene copolymer (30 wt% styrene) and for sulfur vulcanizates of this blend: *(a)* un-cross-lined blend; and *(b–d)* sulfur vulcanization for *(b)* 20, *(c)* 60, and *(d)* 100 min. *cis*-polybutadiene/butadiene-styrene copolymer ratio is 3:7. The arrows indicate the corresponding glass-transition temperatures of individual components. After [14], © John Wiley & Sons, with permission.

can be seen that as a result of vulcanization, the temperature of the low-temperature maximum increases, whereas the second, high-temperature maximum is shifted to a lower temperature. This tendency becomes more distinct with increases in the degree of cross-linking. When the number of linked chains per cubic centimeter N_c exceeds 4.1×10^{19}, it seems possible to regard this system as pseudohomogeneous (Table 6.9).

Thiuram vulcanization at 143°C and radiation vulcanization at 143 and 160°C also lead to homogenization, whereas radiation vulcanization at lower temperatures is less effective (Table 6.9). When vulcanization is performed in two stages (radiation vulcanization at room temperature followed by sulfur or thiuram vulcanization at 143°C), heterogeneity of vulcanizates is determined by the density of links formed in the first stage during irradiation. A pseudohomogeneous structure can be formed only if N_c induced by irradiation does not essentially exceed 0.8×10^{19} (see Section 6.4.3.3). Otherwise, interpenetration of rubbers in the second stage of vulcanization at 143°C becomes difficult. As a result, for irradiation doses higher than 5–10 Mrad, both radiation–thiuram and radiation–sulfur vulcanizates are heterogeneous. If the sequence of vulcanization steps is changed (sulfur or thiuram vulcanization at 143°C preceding irradiation), all the samples are pseudohomogeneous.

Hence a pseudohomogeneous product is formed when vulcanization is carried out at high temperatures and the degree of cross-linking is high enough to prevent phase

Table 6.9. Glass-transition temperatures and the degree of cross-linking of *cis*-polybutadiene–butadiene-styrene copolymer (30 wt% styrene) mixtures

Sample	Glass-transition temperatures defined by radiothermoluminescence (K)	Degree of cross-linking defined by equilibrium stress $N_c \times 10^{-19}$ (cm^{-3})
Un-cross-linked mixture	175, 220	—
Sulfur vulcanizate:		
20 min/143°C	175, 215	1.8
60 min/143°C	190, 210	4.1
100 min/143°C	190, 207	6.3
Thiuram vulcanizate:		
5 min/143°C	187, 204	6.1
10 min/143°C	187, 204	7.0
30 min/143°C	187, 204	7.5
Radiation vulcanizate:		
30 Mrad/20°C	177, 220	3.4
15 Mrad/100°C	180, 218	3.3
10 Mrad/143°C	183, 210	3.0
8 Mrad/160°C	183, 208	3.1
Radiation-thiuram vulcanizate:		
1.5 Mrad/20°C–30 min/143°C	185, 206	8.5
5 Mrad/20°C–30 min/143°C	184, 210	9.1
50 Mrad/20°C–30 min/143°C	183, 215	9.4

separation on subsequent cooling. During heating, the rubbers interpenetrate by mutual diffusion and, after a short period, form an interface layer. Since the interface layer thickness increases with temperature, the size of the region in which the different molecules are co-cross-linked also increases, and cross-linking at higher temperatures results in a vulcanizate with a greater phase homogeneity. Thus, depending on vulcanization conditions and particularly on vulcanization temperature, it is possible to obtain vulcanizates with various degrees of phase heterogeneity.

6.4.3.6 Covulcanization of Elastomers with Polyfunctional Unsaturated Compounds

The cross-linking of rubbers by radiation is known to be accelerated by polyfunctional unsaturated compounds (PFUC), which improve the properties of vulcanizates [157]. Radiation vulcanization produces a steric graft polymerization of the PFUC in the rubber matrix in which the hardened PFUC forms the steric-network joints [158]. The efficiency of radiation cross-linking due to the sensitizing action of PFUCs is governed by their distribution in and interaction with the rubber matrices. Analysis of the glow curves gives useful information about the PFUC–rubber interaction. The major characteristics are the parameters of the rubber-phase T_g peak—its position, intensity, and width. When there is no interaction, the position of the T_g peak and its width are the same as in the cross-linked rubber itself, whereas the intensity of this peak as a function of the ratio of the components is linear [159]. The PFUC–rubber interaction, however, leads to significant changes in the parameters of rubber relaxation in the region of T_g. Evaluation of a variety of PFUCs showed that the largest cross-linking velocity is ensured in polar rubbers. In nonpolar rubbers, the PFUC molecules aggregate into large structural formations and their sensitizing efficiency is small.

6.4.4 Copolymers

As underlined in the previous section, the formation of a homogeneous polymer blend is generally thermodynamically unfavorable. In block and graft copolymers, however, different kinds of incompatible polymers are covalently linked to each other. This restricts, to a certain extent, the demixing process, so for the same two polymers with the same molecular weights, phase separation occurs more readily in simple polymer mixtures than in block or graft copolymers. Nevertheless, as in ordinary phase separation, the tendency is for the domains to grow in order to reduce the interfacial free energy per unit volume, and the phase separation commonly ensues even when the two polymer chains are joined together to form a single molecule. Thus one end of a macromolecule is insoluble in the other end, and phases smaller than the macromolecular dimensions are common. The free-energy level of such a heterogeneous system is determined by a balance between the enthalpy and entropy terms consistent with the equilibrium morphology of the system.

It is clearly not rigorous to view the block or graft copolymer structure as pure domains joined to a unperturbed interface. The joints between blocks are preferentially found in the interfacial regions. As a consequence, there is a loss of entropy in two ways. One is due directly to this confinement of the joint degree of freedom. The other has its origin in the vast number of polymer conformations which are not allowed because they produce lower densities near the domain centers than at the outer boundaries. As the microdomains grow, the entropy losses increase until they finally outweigh the decrease in surface free energy, and the thermodynamically stable size is achieved.

One way to relieve the entropy losses associated with joint confinement and the difficulty of filling the domain centers is to broaden the interface and the interpenetration [160]. In the interlayers, the composition changes continuously from one pure component to another, and sublayers of the boundary have properties of compatible mixtures or random copolymers of the appropriate composition. As blocks become shorter, the domains become smaller and more numerous, resulting in an increase in interfacial area. In addition, the interface becomes more diffuse. When the interlayer attains 100% of the polymer volume, block copolymers no longer contain compositionally pure phases, but are visualized as retaining a residual domain structure in which composition fluctuates between ever-narrowing limits as blocks become shorter, approaching homogeneity and the behavior of a random copolymer.

In terms of relaxational behavior, block and graft copolymers with long-chain sequences of different natures are equivalent to heterogeneous polymer blends; i.e., they exhibit two glass transitions coinciding with those of the individual components. Alternatively, when block lengths are decreased to a level approaching the molecular structure of a random copolymer, only one glass transition located on the temperature scale between the glass transitions of the components is observed. In an intermediate case of short blocks and, consequently, small phase dimensions, the influence of the interface is significant, resulting in a broadening and shifting of the glass transitions toward each other.

6.4.4.1   Random Copolymers

The structure and properties of a copolymer depend on its chemical composition and distribution of monomer units. Copolymerization, which provides random distribution of comonomer units along macromolecules, gives a homogeneous, single-glass-transition copolymer. When both homopolymers are amorphous or the ability of the components to crystallize is supressed by copolymerization, the random distribution of monomers leads to a product with a glass-transition temperature intermediate to those of both homopolymers. Variations in copolymer composition cause a gradual shift of the copolymer's glass-transition temperature, and the compositional dependence of the glass-transition temperature is similar to that for homogeneous polymer blends (Table 6.10).

In some cases, however, one of the components may exhibit a strong tendency toward crystallization, especially when its content is high and, consequently, its unperturbed molecular chain sequences are long. Then crystallization disturbs the monotony of the compositional dependence of T_g, since it is governed not only by compositional but by morphologic changes as well.

Copolymers of ethylene with propylene or vinyl acetate and aliphatic nylons were intensively investigated in efforts to separate morphologic effects from the basic pro-

Table 6.10. Relaxation transition temperatures of dimethylsiloxane (DMS)–methyltrifluoropropylsiloxane (MTFPS) copolymers determined by different methods

DMS:MTFPS (wt%)	Dilatometry		Radiothermoluminescence	
	Secondary transition (K)	Glass transition (K)	Secondary transition (K)	Glass transition (K)
100:0	—	152	128	155
50:50	151	180	152	183
40:60	156	188	150	188
30:70	155	192	162	191.5
20:80	154	197	159	197.5
0:100	156	202	152	204.5

Source: Ref. 186, Pergamon Journals, Inc., with permission.

cesses of isolated methylene sequences. The trends in the transition temperature–composition relation for all polyethylene random copolymers exhibit similar features, and the ethylene–vinyl acetate copolymer can be considered as a typical example (Fig. 6.31a). There are basically two regions where the temperature position of the main relaxation changes differently. As ethylene units are initially introduced into the chain, the copolymer remains amorphous. The material is characterized by a glass-transition temperature which changes in a predictable manner according to one of the standard relations [161]. If crystallization did not occur, then as much higher concentrations of ethylene units are introduced, the glass-transition temperature of pure polyethylene would be approached. However, as the ethylene content increases, the possibility appears for crystallization to take place. When crystallization intervenes, relaxation is not any longer associated with the liquid-like mobility which manifests itself in completely amorphous polymers and is dependent on copolymer composition. The transition temperature remains practically invariant over a broad composition range despite the fact that there are large changes in the relative proportions of noncrystalline material. For instance, for ethylene–vinyl acetate copolymers over the range of 5–28% by weight of vinyl acetate, the enthalpy-determined level of crystallinity varies from 39 to 12% [54].

Two qualitative explanations have been offered to account for the unique transition temperature–composition relation. According to Nielsen [162], when a small amount of ethylene is added to pure polyvinyl acetate, there is a rapid reduction in the temperature of the glass peak. It is to be expected that in the first approximation the glass relaxation temperature will follow the Gordon-Taylor equation [see Eq. (6.12)]. At approximately 40% ethylene, crystallization occurs. Thus, for ethylene concentrations above 40%, the actual ethylene concentration in the amorphous regions is lower than the gross concentration. As a result, the glass-transition temperature remains constant from 40 to 100% ethylene. Reding *et al.* reject this explanation [163]. According to these authors, a copolymer of ethylene and a vinyl monomer must be considered as a mixture of groups containing a few linear CH_2 units and structures containing a tertiary carbon atom, both of which relax independently. At

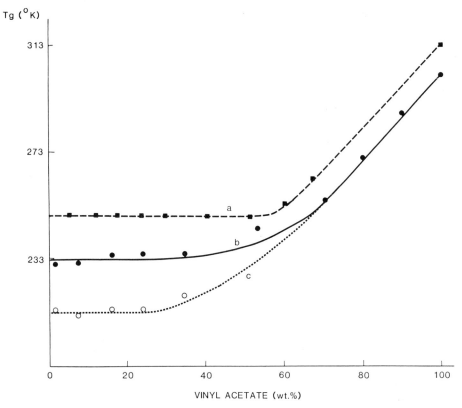

Fig. 6.31. Compositional dependence of glass relaxation for ethylene–vinyl acetate random copolymers: *(a)* mechanical losses at 0.2–1.2 Hz [162]; and *(b)* and *(c)* radiothermoluminescence data for quenched samples; sample cooling rate is, respectively, 60 degrees/second *(b)* and 225 degrees/second *(c)* [86,165].

high-ethylene compositions, the CH_2 groups relax at the γ relaxation ($-110°C$ at ~1 cps, as evaluated by mechanical spectroscopy), and the tertiary carbon structures relax at the β relaxation (-10 to $-50°C$ at ~1 cps depending on the nature of the side group). When the ethylene composition decreases, the magnitude of the relaxation decreases, but the temperature of the relaxation does not change. This occurs because the CH_2 units relax independently of the rest of the chain. When the CH_2 content is less than about 40 wt%, the tertiary carbon structures no longer relax independently of the rest of the chain because of the interaction of side groups. The β relaxation thus rises from the region of $-10°C$, finally reaching a value equal to the T_g value of the vinyl homopolymer. According to Reding *et al.*, in the range 0–65 wt% vinyl acetate, the temperature of the β relaxation changes very little because the pendent groups are isolated from each other whether or not the copolymers are crystalline. This hypothesis was proposed in order to explain the observation that in random ethylene–vinyl acetate copolymers the last vestige of crystallinity disappears at about 45 wt% vinyl acetate, whereas the upward move of the β relaxation occurs at about 65 wt% vinyl acetate. However, theoretically, crystallinity can develop in random copolymers even if they are found to be amorphous at ambient temperatures [164]. Furthermore, Nielsen found that a 43 wt% ethylene–vinyl acetate copolymer had 8% crystallinity at 30°C [162].

The radiothermoluminescence results obtained by Nikolskii *et al.* [86,165] also suggest that Nielsen's hypothesis, which emphasizes the influence of crystallinity on relaxation in copolymers, should be preferred. First of all, the increase in ethylene content from 50 to 100 wt% was found to be accompanied by a gradual widening of the main thermoluminescence maximum around 230 K, its position being practically unchanged (Fig. 6.31*b*). Second, evaluation of thin (25 μm) copolymer films slowly cooled from the melt and rapidly quenched in liquid nitrogen proved that quenching results in a decrease in the transition value of almost 20 K for copolymers with a high ethylene content (Fig. 6.31*c*). The existence of a small horizontal region in the curve is likely associated with the impossibility of obtaining amorphous samples of copolymers with an ethylene content exceeding 80 wt% by utilized quenching procedure.

It has been explicitly pointed out that interpretation of the glass relaxation in ethylene copolymers is complicated by crystallinity [161,166]. Its interpretation as a conventional glass temperature is, therefore, not at all apparent. According to Popli and Mandelkern [161], the glass transition manifests itself only in completely amorphous copolymers, whereas in semicrystalline copolymers, as in polyethylene homopolymers, there is a β relaxation associated with the interfacial regions. However, assignment of the transition around 240 K to the interfacial regions leaves open the question of location of the transition which can be attributed to the interzonal regions. On the other hand, interpretation of this transition as the upper glass-transition temperature $T_g(U)$ in the sense proposed by Boyer [25] seems to be adequate. This is substantiated by the sensitivity of the transition temperature to the conditions of sample preparation observed by means of radiothermoluminescence [86].

It is noteworthy that the extent of the interfacial region is greatly enhanced by random copolymerization [54]. At the same time, the radiothermoluminescence evaluation of a variety of polyethylene copolymers showed that the major change in the glow curves which accompanies introduction of a comonomer in the low-density polyethylene chain is an increase in the intensity of the transition at 178 K, indicating its interfacial origin [167]. Thus this observation is also consistent with assignment of the transition at 240 K to relaxation in interzonal rather than interfacial regions.

It has recently been observed that nearly all copolymerizations involve some degree of phase separation [168]. From this perspective, interesting results were obtained in a comparative study of four butadiene–styrene copolymers with the same styrene content—25 wt% [169]. All materials were characterized as random copolymers by their manufacturers (Table 6.11). First of all, one notices differences in the position of the T_g peak (Fig. 6.32). Low T_g values for Duradene and Taphdene are due to low vinyl content, whereas high T_g for Europrene is most probable caused by a large *trans*-1,4 content. Along with the differences in position of the T_g maxima, each of the glow curves has a different shape. Only Duradene has a sharp, narrow T_g maximum, which is indicative of its structural uniformity. Although Europrene exhibits a strong secondary relaxation (broad maximum centered at ~160 K), sharp

Table 6.11. Structural characteristics of various butadiene-styrene copolymers (25 wt% styrene)

| Trade name | Concentration of cis, trans, and vinyl structures (%) | | | T_g(K) |
	1,4-*cis*	1,4-*trans*	vinyl	
Duradene	44	46	10	198
Taphdene-2000	42	46	12	205
Solprene-1204	28	38	34	217
Europrene-1500	8	74	18	220

Source: Ref. 169.

decay of the glow after passing the T_g peak points to uniform (random) distribution of monomer units. As opposed to Duradene and Europrene, descending portions of the T_g maxima for Taphdene and Solprene are broader. Along with the main maximum, Solprene also exhibits a purely resolved maximum in 180–200 K temperature interval. Thus some of the macromolecular sequences are enreached in butadiene and some in styrene, which leads to phase separation. This conclusion was confirmed when the amount of styrene units forming blocks in copolymer molecules was deter-

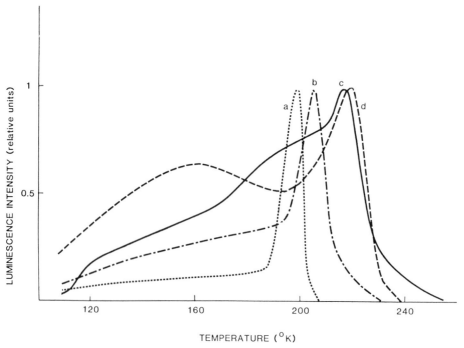

Fig. 6.32. Glow curves of *(a)* Duradene, *(b)* Taphdene-2000, *(c)* Solprene-1204, and *(d)* Europrene-1500 [169].

mined according to the Kolthoff method [170]. This value for Duradene was found to be almost three times smaller than that for Solprene, 1.6 and 4.6, respectively [171].

6.4.4.2 Phase Separation in Block Copolymers

Although the effect of phase separation can be noticed even in some of the "random" copolymers, it is pronounced to a much larger extent in block copolymers. Figure 6.33 shows the glow curve of a styrene-butadiene-sryrene (SBS) block copolymer (28% styrene, $Mw = 10^4/5 \times 10^4/10^4$) along with the glow curves of polystyrene and polybutadiene [140]. The latter two polymers were similar in microstructure and molecular weight to the polybutadiene and polystyrene blocks of the block copolymer. In comparing the T_g peak of polybutadiene with the corresponding maximum on the SBS glow curve, one notices a strongly asymmetrical shape of the latter peak resulting from a broad tail on the high-temperature side of the peak. This suggests the existence of electron traps in the block copolymer that are more stable than

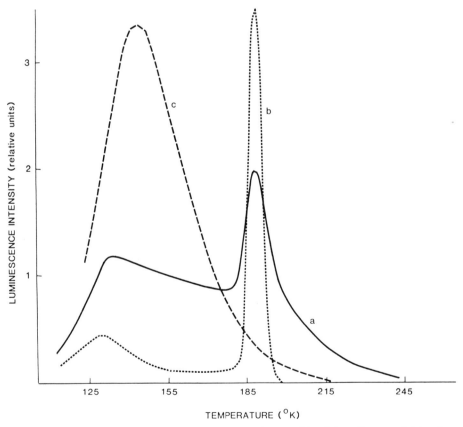

Fig. 6.33. Glow curves for *(a)* an SBS block copolymer, *(b)* polybutadiene, and *(c)* polystyrene. After [140], © 1979, American Chemical Society.

those in polybutadiene. Hence electrons will be freed from their traps and will cause luminescence at $T > T_g$ when all the electron recombination processes have long been completed in the pure polybutadiene phase. As one can see from the glow curve of pure polystyrene, there is no transition occurring near that of the polybutadiene phase in SBS. In fact, appreciable local mobility exists in polystyrene at temperatures considerable below its T_g at 353 K, causing complete erosion of all electron traps in the polymer well before long-range motion sets in. The light given off by polystyrene at about 190 K is of relatively low intensity, since much luminescence has already occurred at lower temperatures. The presence of deeper, more stable traps must thus be sought in regions of the block copolymer, which, have T_gs above that of polybutadiene. The only place where this could be in this material is the interface between the polybutadiene and the polystyrene domains. To define the contribution from this region to the overall luminescence given off by the copolymer, the polybutadiene- and polystyrene-related light emission has to be subtracted from the total glow curve (Fig. 6.34). The volume fraction of the interface X^{IF} can be estimated by assuming a linear concentration gradient of polybutadiene and polystyrene in this region and by neglecting the luminescence contribution of polystyrene to the light emission of the interface. X^{IF} calculated with the preceding two assumptions leads to a value of 0.6 for the solvent-cast sample. Figure 6.34 also indicates that X^{IF} is larger for a

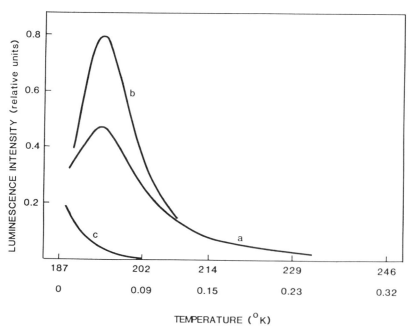

Fig. 6.34. Luminescence contribution of the interface region present in an SBS block copolymer. *(a)* Sample cast from tetrahydrofuran. *(b)* Sample compression molded at 100°C for 10 min. *(c)* Physical blend of 28 wt% polystyrene and 72 wt% polybutadiene. After [140], © 1979, American Chemical Society.

sample molded at 100°C for 10 min compared with that cast from tetrahydrofuran. A comparison of the areas under both curves suggest a ratio $X^{IF}(\text{molded})/X^{IF}(\text{cast}) \simeq 1.1$. Figure 6.34 also clearly demonstrates the larger contribution of the interface in block copolymers as compared to that in physical blends. As block size increases, the influence of the interface diminishes. The block copolymers with very long-chain sequences of individual components can be thought to be similar to corresponding polymer mixtures [172].

6.4.4.3   Radiation-Grafted Systems: Distribution of Grafted Polymer Over the Bulk of the Test Specimen

Graft copolymers are generally prepared by polymerization of a monomer in the presence of a nonpropagating polymer chain. In particular, grafting can be achieved either by irradiation of the polymer in the monomer medium (liquid or vapor) or by irradiation of the polymer alone and its subsequent exposure to the monomer. The latter process is usually referred to as *posteffect grafting*.

Evaluation of grafted materials requires knowledge of whether grafting took place at the surface or in the volume of a substrate. As was shown by Nikolskii *et al.* [173,174], the course of the grafting process and the distribution of grafted polymer can be assessed by means of radiothermoluminescence. If grafting takes place only at the surface of the substrate, the resulting glow curve of the product will be close to the sum of the glow curves of the substrate and homopolymer polymerized during grafting. Uniform grafting which affects not only the surface, but also the inner layers of the substrate is expected to bring about noticeable changes in the shape and intensity of the glow curve of the substrate unless macrophase separation had occurred.

Figure 6.35 shows how grafting of styrene affects the position and intensity of the

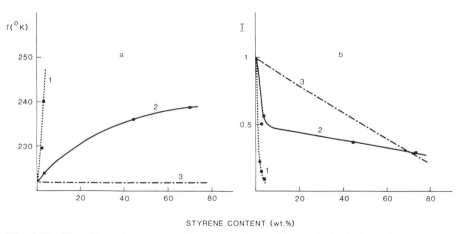

STYRENE CONTENT (wt.%)

Fig. 6.35. The effect of styrene grafting on *(a)* the position and *(b)* the intensity of the main thermoluminescence peak (222 K) in low-density polyethylene (1 = posteffect grafting; 2 = grafting by direct irradiation in the monomer (styrene) vapor; 3 = mechanical blending of polyethylene and polystyrene). After [173], Pergamon Journals, Inc., with permission.

low-density polyethylene luminescence peak at 222 K. It has to be mentioned that the deposition of up to 29 wt% polystyrene on the surface of polyethylene film by immersing polyethylene in a benzene solution of polystyrene practically did not change the position and intensity of the polyethylene transition at 222 K. This was also the case with polyethylene–polystyrene mechanical blends, with the only difference being that the intensity of the transition in question varied according to polyethylene content (Fig. 6.35). These results were not surprising, since at a first approximation, the height of the polyethylene peak at 222 K should be proportional to the fraction of polyethylene in which grafting has not occurred, because both graft copolymer and polystyrene luminesce at this temperature much more weakly than the original polyethylene. In particular, if grafting occurs only in a thin surface layer, the height of the peak will be practically that of polyethylene. Although irradiation of polyethylene up to 10 Mrad shifts its transitions to higher temperatures by several degrees as a result of cross-linking and somewhat decreases the luminescence output, the magnitude of these changes was found to be 5–10 times less than shown in Fig. 6.35 for radiation-grafted materials. Thus it can be concluded that both grafted copolymers cannot be regarded as completely heterogeneous systems, since the polyethylene transition at 222 K was much reduced in amplitude and displaced toward high temperatures. At the same time, grafting by the method of posteffect is more efficient in this respect. Considerable changes occur in this case, even at very low levels of grafting. Just 2–3 wt% of grafted styrene increases the temperature of the thermoluminescence maximum by 15–20 degrees, while the luminescence intensity decreases by a factor of 8–10. This indicates relatively uniform grafting of numerous relatively short polystyrene chains throughout the volume of the substrate.

The probable mechanism of posteffect grafting seems to be as follows: Radicals produced by irradiation rapidly decay when irradiation is stopped. The monomer molecules then freely diffuse into the depth of the film, and grafting is initiated at the crystal surfaces as well as in some transitional zones of different densities which exist between the amorphous and crystal phases, where conditions are most favorable for radical stabilization and monomer penetration. As seen in Fig. 6.35, even 80% of grafting as a result of direct irradiation affects the intensity and position of the polyethylene peak at 222 K to lesser extent than 2% of grafting by posteffect. The independence of the position and intensity of the transition at 222 K from grafted-monomer quantity means that the latter has formed a separate phase. Taking this into consideration, it can be concluded that grafting by direct irradiation, especially at advanced stages of the process, results in nonuniform distribution of grafted material, with some of the grafts forming their own phases. Most probably, in direct graft copolymerization, the large concentration of reactive centers present in polyethylene amorphous regions hinders the monomer penetration in depth. Moreover, the probability of diffusion of monomer molecules into the central part of the film becomes smaller with progressive grafting at the surface.

Results similar to those described above were obtained when styrene was replaced by acrylonitrile, acenaphthylene, and 2,5-methylvinylpyridine monomers [173,174].

6.5 Latex Systems

As distinguished from solid rubbers or plastics, latexes are colloidal suspensions of polymer particles in water. Upon being dried, a film-forming latex is transformed from a milky dispersion into a transparent film in which the contours of the particles observed initially gradually become less distinct.

Because intermolecular interaction is lower at the surface than in the bulk [175], the opportunity for relaxation is much greater at the surface than in the interior of a particle [51]. Thus differences in relaxational behavior have to be expected between latex- and solution-cast films. I am not aware, however, of any methods except radiothermoluminescence [177] which succeed in clearly demonstrating this.

Figure 6.36 shows the glow curves of natural rubber films prepared from latex

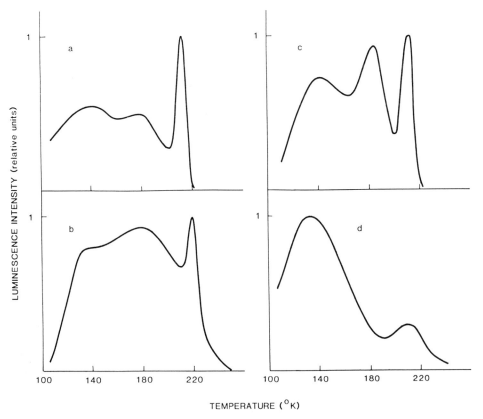

Fig. 6.36. Glow curves for *(a,c)* unvulcanized and *(b,d)* vulcanized natural rubber films cast from *(a,b)* solution and from *(c,d)* natural latex. M_c of vulcanized films was 5500. Composition of the vulcanizing agents (parts per 100 parts of polymer by weight): sulfur, 2.0; ZnO, 3.0; and ethylcymat, 1.0. Vulcanization was performed in an air thermostat at 200°C. After [177], © John Wiley & Sons, with permission.

and a solution in benzene. Both samples have a distinct T_g maximum of the same width at 212.5 K. At the same time, the luminescence peak at 183 K is much more intensive and better resolved in the latex film. This is most probably linked to the presence of a developed interface between the latex globules on which the kinetic mobility of polymer segments is higher than in the interior. The intensity of the peak at 183 K increases with decreasing latex particle size and decreases with the duration of aging, which leads to a gradual disappearance of the interface, or with the addition of a stabilizing mixture (electrolyte solution), which promotes aggregation of the globules.

The final technological step in the processing of both solid rubbers and rubber latexes is vulcanization. There are, however, certain differences between the two cases caused by the peculiar globular structure of latex. Additives introduced into the latex in the form of an aqueous dispersion do not distribute themselves uniformly between the serum and the concentrate in the interglobular spaces, such as are created during the formation of a rubber film from latexes [178]. Two processes take place during the warm-up at vulcanization temperature: (1) vulcanization of a rubber layer on the globule surface which is in direct contact with the vulcanization system, and (2) diffusion of the vulcanization system from the surface layer into the polymer particle. These processes determine the initial rate of vulcanization, which is therefore higher in a latex (due to higher vulcanization system concentration in the interglobular space) than in a solid rubber.

Vulcanization of the natural rubber film prepared from solution does not change the shape of the glow curve substantially (Fig. 6.36). The peak around 183 K remains. The position of the peak in the T_g region is shifted to higher temperatures, and it becomes wider, which is typical for vulcanizates of hard elastomers. Crosslinking, by restricting certain types of motion, causes a spread in the distribution of relaxation times characterizing the process. A glow peak thereby covers a greater temperature range, the spread being in the high-temperature direction. Differently from the solution film, the glow curve of the latex film changes considerable after vulcanization. The maximum at 183 K disappears. Unfreezing of segmental mobility in the T_g region begins at lower temperatures in the vulcanized film than in the nonvulcanized film. The luminescence maximum in the T_g region broadens considerably without changing its position along the temperature axis.

These distinctions are apparently connected with the different structures of films prepared from solution and latex films. The homogeneous structure of films from the solution, as opposed to latex films, as well as the more uniform distribution of vulcanization agents in them, creates conditions for a relatively identical bonding of the polymer throughout the entire volume. Hence the vulcanization of films from solution does not substantially affect their phase homogeneity. Typical of a latex film, however, is a globular structure with definite interstices between the globules. The vulcanizing ingredients are concentrated in these interstices in the immediate vicinity of the mobile segments on the globule surface. This, in turn, creates conditions under which vulcanization affects mainly the surface layers of the globules and thus leads to intensive interglobular bonding. As a result, the free interglobular surface decreases and the relaxation transition at 183 K vanishes. Nevertheless, a somewhat greater mobility of the outer globular layers even after cross-linking is apparent in the glow curves from the fact that the increased segmental mobility in the T_g region begins at lower temperatures in vulcanized latex film than in nonvulcanized film. An abnormal broadening of the maximum at T_g, owing to the vulcanization of latex film, results directly from its microheterogeneity and nonstatistical distribution of cross-links.

Because vulcanization of globules proceeds mainly on their surface, the structure of the surface layers, particularly those containing reactive groups, is important. The vulcanization of latexes does not necessarily result in coalescence of globules into one coherent, continuous mass, but it may, to a certain extent, promote just the opposite effect. It was noted that resolution of the transitions manifesting surface mobility in butadiene–acrylonitrile latex film improved as a result of vulcanization [11]. Following vulcanization, a weak shoulder observed in the glow curve of non-vulcanized film around 205 K transformed into a distinct, well-resolved maximum. It is well known that the distribution of monomer units along the macromolecules in butadiene–acrylonitrile copolymers is not regular, leading to heterogeneity. Acrylo-nitrile, the more polar monomer, concentrates during emulsion polymerization on the surfaces of the globules because of the affinity of the nitrile groups to water [179], whereas the inner parts of the globules contain a greater number of nonpolar buta-diene units. It can, however, be conjectured that the presence on the globule surface of even a small number of places for the bonding of sulfur (double bonds in the butadiene units) results in a high degree of structuring on the part of the globule surfaces where the butadiene links are localized. The resultant shrinkage promotes a decrease of contact between the globule interfaces along which no interglobular bonding has taken place and improves the resolution of the transitions revealing surface relax-ation.

Thus use of the radiothermoluminescence method makes it possible to establish that there are substantial differences in the relaxation spectra of nonvulcanized and, especially, of vulcanized films prepared from solution and latex films and to connect these differences to distinctions in structure.

6.6   Oriented Systems

Orientation of semicrystalline polymers results in a broadening and a shift of the thermoluminescence maxima toward high temperatures [180]. In the region of Hoo-kean deformation, the shift is relatively small—5–7 degrees. It sharply increases when yield stress is reached and then levels off in the region of forced-rubber-like elasticity. Finally, it increases again at large strains, i.e., when extension is due to extension of straightened macromolecules. The magnitude of the shift in transitions depends not only on the extent of deformation, but also on the temperature at which the material was strained [181]. At low temperatures, where relaxation is suppressed, orientation and, consequently, the upward shift of the transition temperatures increase with an increase in temperature. At high temperatures, relaxation is a determinant factor, and the opposite effect is observed. Thus the shift in transitions is defined not by the absolute value of deformation, but by the level of unrelaxed mechanical stress

accumulated in the sample. For polyethylene, the temperature of deformation which
gives the largest shift in transition temperatures was found to be 300 K [181]. The
frequency of polyethylene segmental relaxation at this temperature is 10^4–10^5 Hz
[102]. At the same time, the stressed state in an oriented sample remains much
longer. Hence fixation of the stress is due to macromolecular formations much larger
than segments. When polyethylene first extended at 300 K and then freed from the
action of the external forces is heated up to 360 K, it regains its initial shape and
exhibits the glow curve characteristic for nonoriented material.

Interesting results were obtained when, instead of orientation and subsequent irradiation at liquid nitrogen temperature according to the standard radiothermoluminescence procedure, these two operations were switched around [182]. The original
polyethylene film exhibited two major thermoluminescence maxima at 181 and 232
K which arise from interfacial and interzonal mobilities, respectively (see Table 6.6),
and can be identified in terms proposed by Boyer [25] with $T_g(L)$ and $T_g(U)$ transitions. When the stress applied to the pre-irradiated polyethylene film was greater than
270 kg cm^{-2}, there was a new flash on the glow curve (called the δ peak by the
authors) which appeared at the moment of film rupture (Fig. 6.37). For stresses less
than 270 kg cm^{-2}, the δ peak was not observed and the rupture occurred when
luminescence practically vanished. At this condition, a shift to lower temperatures of
both thermoluminescence maxima could be noted (Fig. 6.38a,b). For stresses ex-

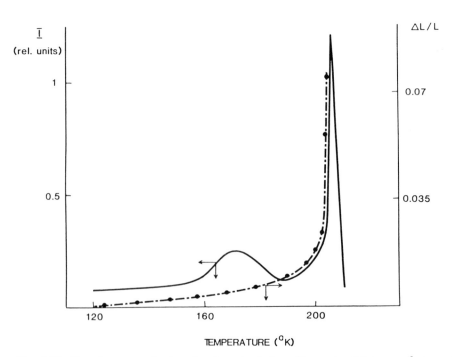

Fig. 6.37. The glow curve for low-density polyethylene film under 400 kg cm^{-2} stress. The
dashed line shows the change in relative elongation with temperature [182].

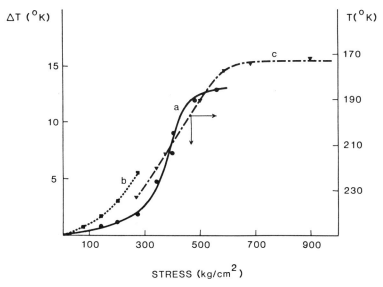

Fig. 6.38. Shift in the positions of the transitions *(a)* $T_g(L)$, and *(b)* $T_g(U)$ and *(c)* the position of the δ maximum as a function of applied stress [182].

ceeding 600 kg cm^{-2}, the rupture was accompanied by a neglegible elongation and the temperature of the δ peak reached a constant value ($\sim$170 K) which lies in the region of $T_g(L)$ relaxation and can probably be identified with the polyethylene brittleness temperature (Fig. 6.38c). This indicates the connection between the onset of ductility and the polyethylene $T_g(L)$ relaxation.

6.7 Filled Systems

According to experimental evidence existing in the literature, the introduction of a reinforcing filler into a polymer matrix causes T_g of the latter to increase, unless imperfections counterbalance the reinforcing effect [183]. The results have been explained quantitatively in terms of different conformational properties and restricted molecular mobility of a polymer near the filler interface.

Only a few filled systems have been evaluated by the radiothermoluminescence method. An increase in transition temperature has been reported by Vonsyatskii and Boyarskii [184] for a polyethylene–aerosil system (Fig. 6.39). The possibility of homogenization of the blend of polypropylene with ethylene-propylene copolymer under high pressure and shearing strain has been studied by Zhorin et al. [185], and it was shown that introduction into the blend of an active filler (aerosil, asbestos, or carbon black) allows a two- to fourfold reduction in the minimum pressure needed for the formation of a stable homogeneous system.

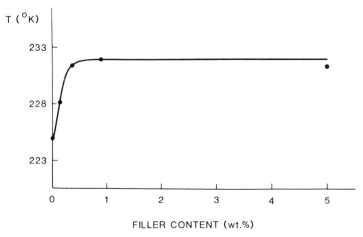

Fig. 6.39. Transition temperature versus aerosil content for a high-density polyethylene [184].

References

1. Ferry, J. D.: Viscoelastic Properties of Polymers. Wiley, New York, 1980
2. Bueche, F.: Physical Properties of Polymers. Interscience, New York, 1962
3. Aulov, V. A., Lednev, I. K., Perekupka, A. G., *et al.:* Polym. Sci. USSR *25*, 1081 (1983)
4. Partridge, R. H.: The Radiation Chemistry of Macromolecules, vol. 1. (ed. Dole, M.). Academic Press, New York, 1972
5. Zlatkevich, L. Y., Crabb, N. T.: J. Polym. Sci. Polym. Phys. Ed. *19*, 1177 (1981)
6. Nielsen, L. E.: J. Macromol. Sci. Rev. Macromol. Chem. *C3*, 69 (1969)
7. Fox, T. G., Loshaek, S.: J. Polym. Sci. *15*, 371 (1955)
8. DiMarzio, E. A.: J. Res. Nat. Bur. Std. *68A*, 611 (1964)
9. DiBenedetto, A. T.: Unpublished results
10. Nikolskii, V.: Sov. Sci. Rev. *7*, 77 (1972)
11. Zlatkevich, L. Y.: Rubber Chem. Technol. *49*, 179 (1976)
12. Alfimov, M. V., Nikolskii, V. G.: Polym. Sci. USSR *5*, 477 (1963)
13. Buben, N. Ya., Goldanskii, V. I., Zlatkevich, L. Y., Nikolskii, V. G., Raevskii, V. G.: Proc. Acad. Sci. USSR Phys. Chem. *162*, 386 (1965)
14. Nikolskii, V. G., Zlatkevich, L. Y., Borisov, V. A., Kaplunov, M. Ya.: J. Polym Sci. Polym. Phys. Ed. *12*, 1259 (1974)
15. McCrum, N. G., Read, B. E., Williams, G.: Anelastic and Dielectric Effects in Polymeric Solids. Wiley, New York, 1967
16. McCall, D. W.: Nat. Bur. Stand. (US) Spec. Pub. *301*, 475 (1969)
17. Johari, G. P., Goldstein, M.: J. Chem. Phys. *53*, 2372 (1970)
18. Johari, G. P.: Ann. N.Y. Acad. Sci. *279*, 117 (1976)
19. Alfimov, M. V., Nikolskii, V. G.: Polym. Sci. USSR *5*, 477 (1963)
20. Kozlov, V. T., Ivanov, S. I., Smagin, E. .N.: High Energy Chem. *1*, 350 (1967)
21. Bohm, G. G. A.: J. Polym. Sci. Polym. Phys. Ed. *14*, 437 (1976)
22. Zlatkevich, L. Y., Nikolskii, V. G.: Vysocomol. Soedin. [B] *11*, 810 (1971)
23. Wunderlich, B.: Macromolecular Physics, vol. 1. Academic Press, New York, 1973.

24. Mindiyarov, Kh. G., Zelenev, Yu. V., Bartenev, G. M.: Vysocomol. Soedin. [B] *17*, 527 (1975)
25. Boyer, R. F.: Macromolecules *6*, 288 (1973)
26. Mindiyarov, Kh. G., Zelenev, Yu. V., Bartenev, G. M.: Polym. Sci. USSR *14*, 2737 (1972)
27. Kunizhev, B. E., Belaev, O. Ph., Kuyumdge, E. C., Zelenev, Yu. V.: Vysocomol. Soedin. [B] *23*, 186 (1981)
28. Starkweather, H. W.: Macromolecules *14*, 1277 (1981)
29. Kauzmann, W.: Rev. Mod. Phys. *14*, 12 (1942)
30. Shen, M. C., Eisenberg, A.: Prog. Solid State Chem. *3*, 407 (1967)
31. Knappe, W., Voigt, G., Zyball, A.: Colloid Polym. Sci. *252*, 673 (1974)
32. Nikolskii, V. G., Burkov, G. I.: High Energy Chem. *5*, 373 (1971)
33. Aharoni, S. M.: J. Appl. Polym. Sci. *16*, 3275 (1972)
34. Fleming, R. J.: J. Polym. Sci., part A2, *6*, 1283 (1968)
35. Rafikov, S. R., Korobeinikova, V. N., *et al.:* Polym. Sci. USSR *20*, 863 (1978)
36. Charlesby, A., Partridge, R. H.: Proc. R. Soc. *A271*, 170 (1963)
37. Rozman, I. M.: Bull. Acad. Sci. USSR Phys. Ser. *22*, 48 (1958)
38. Pender, L. F., Fleming, R. J.: J. Phys. [C] *10*, 1571 (1977)
39. Partridge, R. H.: Ph.D. Thesis, Univ. of London (1964)
40. Nikolskii, V. G.: Proc. Acad. Sci. USSR Phys. Chem. *176*, 663 (1967)
41. Partridge, R. H., Charlesby, A.: J. Polym. Sci. Polym. Lett. Ed. *1*, 439 (1963)
42. Boustead, I.: Nature *225*, 846 (1970)
43. Hashimoto, T., Sakai, T., Iguchi, M.: J. Phys. [D] *12*, 1567 (1979)
44. Enn, J. B., Shimha, R.: J. Macromol. Sci. Phys. *B13*, 25 (1977)
45. Hashimoto, T., Ogita, K., Umemoto, S.: J. Polym. Sci. Polym. Phys. Ed. *21*, 1347 (1983)
46. Miller, R. L.: Encyclopedia of Polymer Science and Technology, vol 4. Interscience, New York, 1966
47. Boyer, R. G.: Br. Polym. J. *14*, 163 (1982)
48. Hoffman, J. D., Williams, G., Passaglia, E.: J. Polym. Sci. [C] Polym. Symp., No. 14, 173 (1966)
49. Eby, R. K., Sinnott, K. M.: J. Appl. Phys. *32*, 1765 (1961)
50. Scott, A. H., Scheiber, D. J., Curtis, A. J., Lauritzen, J. I., Hoffman, J. D.: J. Res. Nat. Bur. Std. *66A*, 269 (1962)
51. Boyer, R. F.: Polymer *17*, 996 (1976)
52. Bergmann, V. K., Nawotki, K.: Kolloid-Z. *219*, 131 (1967)
53. Glotin, M., Domszy, R., Mandelkern, L.: J. Polym. Sci. Polym. Phys. Ed. *21*, 285 (1983)
54. Glotin, M., Mandelkern, L.: Coll. Polym. Sci. *260*, 182 (1982)
55. Stehling, F.C., Mandelkern, L.: Macromolecules *3*, 242 (1970)
56. Davis, G.T., Eby, R.K.: J. Polym. Phys. *44*, 4274 (1973)
57. Ashcraft, C.R., Boyd, R.H.: J. Polym. Sci. Polym. Phys. Ed. *14*, 2153 (1976)
58. Sinnott, K.M.: J. Polym. Sci. [C] Polym. Symp., No. 14, 141 (1966)
59. Illers, K.H.: Rheol. Acta *3*, 202 (1964)
60. Pechhold, W., Blasenbrey, S., Woerner, S.: Kolloid-Z. *189*, 14 (1963)
61. Saito, N., Okano, K., Iwayanagi, S., Hideshima, T.: Solid State Phys. *14*, 343 (1963)
62. Schatzki, T.F.: J. Polym. Sci. *57*, 496 (1962)
63. Schmieder, K., Wolf, K.: Kolloid-Z., *134*, 149 (1953)
64. Oakes, W.G., Robinson, D.W.: J. Polym. Sci. *14*, 505 (1954)
65. Willbourn, A.H.: Trans. Faraday Soc. *54*, 717 (1958)

66. Lam, R., Geil, P.H.: Polym. Bull. *1*, 127 (1978)
67. Lee, S., Simha, R.: Macromolecules *7*, 909 (1974)
68. Rathje, J., Ruland, W.: Colloid. Polym. Sci. *254*, 358 (1976)
69. Moore, R.S., Shiro, M.: J. Polym. Sci. [C] *5*, 163 (1963)
70. Pechhold, W., Eisele, V., Knauss, G.: Kolloid-Z. *196*, 27 (1964)
71. Illers, V.K.H.: Kolloid-Z. Z. Polym. *251*, 394 (1973)
72. Cooper, J.W., McCrum, N.G.: J. Mater. Sci. *7*, 1221 (1972)
73. Kitamaru, R., Horii, F., Hyon, S.H.: J. Polym. Sci. Polym Phys. Ed. *15*, 821 (1977)
74. Kitamaru, R., Horii, F., Hyon, S.H.: Am. Chem. Soc. Polym. Prep. *17*, 546 (1976)
75. Kitamaru, R., Horri, F.: Adv. Polym. Sci. *26*, 137 (1978)
76. Popli, R., Glotin, M., Mandelkern, L.: J. Polym. Sci. Polym. Phys. Ed. *22*, 407 (1984)
77. Boustead, I., Charlesby, A.: Proc. R. Soc. *A316*, 291 (1970)
78. Boustead, I.: Proc. R. Soc. *A319*, 237 (1970)
79. Walter, E.R., Reding, F.P.: J. Polym. Sci. *21*, 561 (1956)
80. Spevacek, J.: Polymer *19*, 1149 (1978)
81. Kobayashi, K., Nagasawa, T.: J. Macromol. Sci. *B4*, 331 (1970)
82. Sumita, M., Miyasaka, K., Ishikawa, K.: J. Polym. Sci. Polym. Phys. Ed. *15*, 837 (1977)
83. Balta Calleja, F.J., Hosemann, R.: J. Polym. Sci. Polym. Phys. Ed. *18*, 1159 (1980)
84. Philipov, V.V., Nikolskii, V.G.: Vysokomol. Soedin. *B24*, 372 (1982)
85. Buben, N.Ya., Goldanskii, V.I., Zlatkevich, L.Yu., *et al.:* Proc. Acad. Sci. USSR Phys. Chem. *178*, 30 (1968)
86. Nikolskii, V.G., Plate, I.V., Fazlyyev, F.A., *et al.:* Polym. Sci. USSR *25*, 2750 (1983)
87. Osintseva, L.A., Zlatkevich, L.Yu., *et al.:* Polym. Sci. USSR *16*, 394 (1974)
88. Aulov, V.A., Sukhov, F.F., Slovokhotova, N.A.: Vysokomol. Soedin. *B15*, 615 (1973)
89. Reneker, D.H.: J. Polym. Sci. *59*, 539 (1962)
90. Boustead, I., George, T.J.: J. Polym. Sci. Polym. Phys. Ed. *10*, 2101 (1972)
91. Illers, K.H.: Kolloid-Z. Z. Polym. *231*, 622 (1969)
92. Blake, A.E., Charlesby, A., Randle, K.J.: J. Phys. [D] *7*, 759 (1974)
93. Nakamura, S., Ieda, M.: J. Appl. Phys. *48*, 179 (1977)
94. Hashimoto, T., Ogita, K., Umemoto, S., Sakai, T.: J. Polym. Sci. Polym. Phys. Ed. *21*, 1347 (1983)
95. Nakamura, S., Sawa, G.: J. Appl. Phys. *48*, 3626 (1977)
96. Miller, R.L.: Polymer *1*, 135 (1960)
97. Hosemann, R.: Acta Cryst. *4*, 520 (1951)
98. Natta, G.: Soc. Plast. Eng. J. *15*, 368 (1959)
99. Woodward, A.E., Sawer, J.A.: Physics and Chemistry of the Organic Solid State. vol 2. (ed. Fox, D.). Interscience, New York, 1965
100. Karasz, F.E., MacKnight, W.J.: Macromolecules *1*, 537 (1968)
101. Burfield, D.R., Doi, Y.: Macromolecules *16*, 702 (1983)
102. Kargin, V.A., Andrianova, G.P., Kardash, G.G.: Polym. Sci. USSR *9*, 289 (1967)
103. Ke, B.: J. Polym. Sci. [B] *1*, 167 (1963)
104. Draiu, K.F., Murphy, W.R., Otterburn, M.S.: Polymer *24*, 553 (1983)
105. Blackadder, D.A., Le Poidevin, G.J.: Polymer *17*, 769 (1976)
106. Patterson, D.: J. Paint Technol. *45*, 37 (1973)
107. Zlatkevich, L.: Unpublished results
108. Nakajima, A., Fujiwara, H.: Bull. Chem. Soc. Japan *37*, 909 (1964)
109. Nishioka, A., Koike, Y., Owaki, M., Naraba, T., Kato, Y.: J. Phys. Soc. Japan *15*, 416 (1960)
110. Nikolskii, V.G., Zlatkevich, L.Yu., Osintseva, L.A., Konstantinopolskaya, M.B.: Polym. Sci. USSR *16*, 3209 (1974)

111. Beck, B.L., Hiltz, A.A., Knox, J.R.: SPE Trans. *3*, 279 (1963)
112. Wada, T., Hotta, T., Susuki, R.: J. Polym. Sci. [C] *23*, 583 (1968)
113. Keith, K.D., Padden, F.I.: J. Appl. Phys. *35*, 1270 (1964)
114. Wilkinson, R.W., Dole, M.: J. Polym. Sci *58*, 1089 (1962)
115. Passaglia, E., Martin, G.M.: J. Res. Nat. Bur. Std. *A68*, 519 (1964)
116. Rice, M.H., McQueen, R.G., Walsh, J.M.: Solid State Physics: Advances in Research and Applications, Vol. 6. (ed. Seitz, F., Turnbull, D.). Academic Press, New York, 1958
117. Styrikovitch, N.M., Adadurov, G.A., Gustov, V.W., *et al.:* J. Polym. Sci Polym. Lett. Ed. *13*, 641 (1975)
118. Petermann, J., Schultz, J.M.: J. Mater. Sci. *13*, 2188 (1978)
119. Zubov, Yu.A., Selihova, V.J., Kargin, V.A.: Polym. Sci. USSR *9*, 394 (1967)
120. Sperati, C.A., Starkweather, H.W.: Adv. Polym. Sci. *2*, 465 (1961)
121. Mele, A., Site, A.D., Bettinali, C., Domenico, A.D.: J. Chem. Phys. *49*, 3297 (1968)
122. Tomita, A.: J. Phys. Soc. Japan *28*, 731 (1970)
123. Rigby, H.A., Bunn, C.W.: Nature *164*, 583 (1949)
124. Furukawa, G.T., McCoskey, R.E., King, G.J.: J. Res. Nat. Bur. Std. *49*, 273 (1952)
125. Rigby, H.A., Bunn, C.W.: Nature *164*, 583 (1949)
126. Hyndman, D., Origlio, G.F.: J. Appl. Phys. *31*, 1849 (1960)
127. Nikolskii, V.G., Buben, N.Ya.: Proc. Acad. Sci. USSR Phys. Chem. *134*, 827 (1960)
128. Flory, P.J.: Principles of Polymer Chemistry. Cornell Univ. Press, Ithaca, New York, 1953
129. Kovar, J., Fortelny, I., Bohdanecky, M.: Int. Polym. Sci. Technol. *9*, T/50 (1982)
130. Krause, S.: J. Macromol. Sci. *C7*, 251 (1972)
131. Inoue, T.: Int. Polym. Sci. Technol. *8*, T/65 (1981)
132. Paul, D.R., Barlow, J.W.: Macromol. Sci. Rev. Macromol. Chem. *C18*, 109 (1980)
133. Allen, G., Gee, G., Nicholson, J.P.: Polymer *2*, 8 (1961)
134. Derham, K., Goldsbrough, J., Gordon, M.: Pure Appl. Chem. *38*, 97 (1974)
135. Bohm, G.G.A., Lucas, K.R., Mayes, W.G.: Rubber Chem Technol. *50*, 714 (1977)
136. Buben, N.Ya., Goldanskii, V.I., Zlatkevich, L.Yu., *et al.:* Polym. Sci. USSR *9*, 2575 (1967)
137. Sperling, L.H., Huelck, V., Thomas, D.A.: Macromolecules *4*, 435 (1972)
138. Stoelting, J., Karasz, F.E., MacKnight, W.J.: Polym. Eng. Sci. *10*, 133 (1970)
139. Kaplan, D.S.: J. Appl. Polym. Sci. *20*, 2615 (1976)
140. Bohm, G.G.A., Lucas, K.R.: Adv. Chem. Ser. *174*, 227 (1979)
141. Williams, F.: The Radiation Chemistry of Macromolecules, vol. 1. (ed. Dole, M.). Academic Press, New York, 1972
142. Zlatkevich, L.Yu., Nikolskii, V.G.: Rubber Chem. Technol. *46*, 1210 (1973)
143. Melnikova, O.L., Kuleznev, V.N., Aulov, V.A., Klykova, V.D.: Vysokomol. Soedin. [B] *18*, 903 (1976)
144. Pestov, S.S., Kuleznev, V.N., Shershnev, V.A.: Kolloid-Z *40*, 581 (1978)
145. Scott, R.L.: J. Polym. Sci. *9*, 423 (1952)
146. Bakeev, N.Ph., Zharikova, Z.Ph., Malinskii, U.M., Izumnikov, A.L.: Vysokomol. Soed. *B19*, 832 (1977)
147. Shershnev, V.A., Pestov, S.S.: Kauch. Resina 9, 11 (1979)
148. Gordon, M., Taylor, J.S.: J. Appl. Chem. *2*, 493 (1952)
149. Fox, T.G.: Bull. Am. Phys. Soc. *1*, 123 (1956)
150. Zlatkevich, L.Yu., Nikolskii, V.G., Raevskii, V.G.: Vysokomol. Soed. *B11*, 310 (1969)
151. Zlatkevich, L.Yu., Nikolskii, V.G., Raevskii, V.G.: Proc. Acad. Sci. USSR Phys. Chem. *176*, 748 (1967)
152. Flory, P.J., Rehner, J.: J. Chem. Phys. *11*, 521 (1943)

153. Zhorin, V.A., Mironov, N.A., et al.: Int. Polym. Sci. Technol. 9, T/21 (1982)
154. Zhorin, V.A., Mironov, N.A., Nikolskii, V.G., Enikolopyan, N.S.: Polym. Sci. USSR 22, 440 (1980)
155. Zhorin, V.A., Mironov, N.A., Nikolskii, V.G., Enikolopyan, N.S.: Proc. Acad. Sci. USSR Phys. Chem. 244, 112 (1979)
156. Gaylord, N.G.: Adv. Chem. Ser. 142, 76 (1975)
157. Kuzminskii, A.S., Fedoseyeva, T.S., Makhlis, F.A.: Zhur. Vsesoyuz. Khim. Obsh. im. D.I. Mendeleeva 18, 285 (1973)
158. Makhlis, F.A., Nikitin, L.Ya., Kuzminskii, A.S.: Polym. Sci. USSR 17, 198 (1975)
159. Khazimkhametov, F.F., Makhlis, F.A., Nikolskii, V.G., Zelenev, Yu.V.: Polym. Sci. USSR 19, 2606 (1977)
160. Helfand, E.: Rubber Chem Technol. 49, 237 (1976)
161. Popli, R., Mandelkern, L.: Poly. Bull. 9, 260 (1983)
162. Nielsen, L.E.: J. Polym. Sci. 42, 357 (1960)
163. Reding, F.P., Faucher, J.A., Whitman, R.D.: J. Polym. Sci. 57, 483 (1962)
164. Flory, P.J.: Trans. Faraday Soc. 51, 848 (1955)
165. Nikolskii, V.G., Terteryan, R.A., Livshiz, S.D., et al.: Vysokomol. Soed. [B] 17, 514 (1975)
166. Flocke, H.A.: Kolloid-Z. 180, 118 (1962)
167. Samoilov, S.M., Aulov, V.A.: Polym. Sci. USSR 18, 1124 (1976)
168. Kollinsky, F., Markert, G.: Multicomponent Polymer Systems. American Chemical Society, Washington D.C., 1971
169. Masagutova, L.V., Zlatkevich, L.Yu., Guseva, V.I., Nikolskii, V.G.: Vysokomol. Soed. [B] 13, 881 (1971)
170. Kolthoff, I., Lee, T.S., Carr, C.W.: J. Polym. Sci. 1, 429 (1946)
171. Nikolskii, V.G.: Pure Appl. Chem. 54, 493 (1982)
172. Nikolskii, V.G., Zlatkevich, L.Yu., Krol, V.A., Petrov, G.N.: Vysokomol. Soed [B] 9, 691 (1968)
173. Nikolskii, V.G., Danilov, Ye.P., Mironov, N.A., Karpov, V.L.: Polym. Sci. USSR 12, 1462 (1970)
174. Nikolskii, V.G., Krasotkina, I.A., Tikhomirova, N.S., Tubasova, I.A.: High Energy Chem. 5, 128 (1971)
175. Vettegren, V.E., Chmel, A.: Vysokomol. Soed. [B] 18, 521 (1976)
176. Fleming, R.J., Pender, L.F.: J. Electrostat. 3, 139 (1977)
177. Zlatkevich, L., Shepelev, M.: J. Polym. Sci. Polym. Phys. Ed. 16, 427 (1978)
178. Duchacek, V.: Int. Polym. Sci. Technol. 9, T/35 (1982)
179. Muroi, S.: J. Appl. Polym. Sci. 10, 713 (1966)
180. Shulyak, A.D., Yerofeyev, V.S. et al.: Polym. Sci. USSR 13, 1234 (1971)
181. Tochin, V.A., Saposhnikov, D.N., Nikolskii, V.G.: Vysokomol. Soed. [B] 12, 609 (1970)
182. Nikolskii, V.G., Saposhnikov, D.N., Tochin, V.A.: Vysokomol Soed. [B] 12, 19 (1970)
183. Theocaris, P.S., Spathis, G.D.: J. Appl. Polym. Sci. 27, 3019 (1982)
184. Vonsyatskii, V.A., Boyarskii, G.Ya.: New Methods of Polymer Investigation (ed. Lipatov, Yu.S.). Nauckova Dumka, Kiev, 1975
185. Zhorin, V.A., Kulakov, V.V., Mironov, N.A., et al.: Polym. Sci. USSR 24, 1081 (1982)
186. Abeliov, Ya.A., Nikolskii, V.G., Kirillov, V.N., Alekseyev, B.F.: Polym. Sci. USSR 20, 2590 (1978)
187. Starkweather, H.W.: J. Macromol. Sci. B2, 781 (1968)

Index